Study Guide/Solutions Manual

to accompany

Genetics: From Genes to Genomes

Leland Hartwell
Fred Hutchinson Cancer Research Center

Lee Hood
University of Washington

Michael Goldberg
Cornell University

Ann Reynolds
University of Washington

Lee Silver
Princeton University

Ruth Veres

Prepared by

Ann Reynolds
University of Washington

Boston   Burr Ridge, IL   Dubuque, IA   Madison, WI   New York   San Francisco   St. Louis
Bangkok   Bogotá   Caracas   Lisbon   London   Madrid
Mexico City   Milan   New Delhi   Seoul   Singapore   Sydney   Taipei   Toronto

McGraw-Hill Higher Education

A Division of The McGraw-Hill Companies

Study Guide/Solutions Manual to accompany
GENETICS: FROM GENES TO GENOMES.

This book is printed on recycled, acid-free paper
containing 10% postconsumer waste.

RECYCLED

1 2 3 4 5 6 7 8 9 0 QPD/QPD 9 0 3 2 1 0 9

ISBN 0-07-029783-5

www.mhhe.com

Table of Contents

1	Mendel's Breakthrough: Patterns, Particles and Principles of Heredity	1
2	Extensions to Mendel: Complexities in Relating Genotype to Phenotype	12
3	The Chromosome Theory of Inheritance	24
4	Linkage, Recombination, and the Mapping of Genes on Chromosomes	33
5	DNA: The Molecule of Heredity Carries, Replicates, and Recombines Information	49
6	Anatomy and Function of a Gene: Dissection through Mutation	55
7	Gene Expression: The Flow of Genetic Information from DNA via RNA to Protein	65
8	DNA at High Resolution: Use of DNA Cloning, PCR, and Hybridization as the Tools of Genetic Analysis	74
9	The Direct Detection of Genotype	85
10	The Mapping and Analysis of Genomes	92
11	The Eukaryotic Chromosome: An Organelle for Packaging and Managing DNA	102
12	Chromosomal Rearrangements and Changes in Chromosome Number Reshape Eukaryotic Genomes	108
13	The Prokaryotic Chromosome: Genetic Analysis in Bacteria	121
14	The Chromosomes of Organelles Outside the Cell Nucleus Exhibit Non-Mendelian Patterns of Inheritance	131
15	Gene Regulation in Prokaryotes	135
16	Gene Regulation in Eukaryotes	143
17	Cell-Cycle Regulation and the Genetics of Cancer	148
18	*Saccharomyces cerivisiae*: Genetic Portrait of Yeast	153
19	*Arabidopsis thaliana*: Genetic Portrait of a Model Plant	158
20	*Caenorhabditis elegans*: Genetic Portrait of a Simple Metazoan	162
21	*Drosophila melanogaster*: Genetic Portrait of a Fruit Fly	167
22	*Mus musculus*: Genetic Portrait of a House Mouse	174
23	The Genetic Analysis of Populations and How They Evolve	179
24	Evolution at the Molecular Level	190

Introduction

Genetics professors agree: the way to learn genetics is through doing many problems. Do as many problems as you can, keeping up with the problems relevant for that reading. Do a little bit each day. Don't wait until the night before an exam to try all the problems or you will be overwhelmed and probably will become very frustrated. You will develop your problem-solving skills by practice.

Many of you will find that genetics requires a set of analytical skills that may not have been called upon in your other biology course. Throughout the course you will have to solve word problems, think experimentally, and make decisions about what information you have learned that is useful in answering a question. Many of the analytical and logic skills that you develop will be broadly applicable in other disciplines. Tests will not have exactly the same questions or wording so it is important that you develop your ability to reason through a problem rather than learning how to solve a particular type of problem.

You will encounter different types of problems throughout the book.

- The classical genetics problems (chapters 1-4; 11-13) are like puzzles. Try to not get frustrated by any one problem and if you do, take a breather and come back to the problem later.

- Experimental types of questions found more often in chapter 5-10, 15-17 on molecular analyses often require a good understanding of a variety of techniques so you can choose an appropriate technique. Make sure you have a good handle on the fundamental techniques such as gel electrophoresis, hybridization principles. When would each be used? How is each done? What materials do you need (cells, DNA, RNA or protein).

- In the genetic portrait chapters you get a chance to integrate your accumulated knowledge and think more like a scientist pursuing a specific line of research using genetic analysis. You will need to draw on all your resources to decide if you need to do crosses, or analyze molecular events or a combination of both?

Solutions have not been provided for the Social and Ethical Issue scenarios because there is no one simple answer. There are many people involved or impacted by each issue (we call these stakeholders) and each has valid concerns and issues that cause them to hold particular viewpoints. It is important to consider each person's likely viewpoints based on their values and their roles in society. These complex ethical issues benefit from discussion and often require compromises on the part of the participants.

How this student guide will help you:

- **Synopsis**   The synopsis is a distillation of what each chapter is about. It is good practice to try to do this for yourself, not only for chapters in the text, but also after a lecture. Try to come up with the main focus of the chapter or lecture and think about the information in the context of what else you have learned.

- **Be prepared to:** This section reminds you of the types of problems and applications that you should be able to answer given the material in the chapter.

- **Problem solving tips:** This includes items to remember and useful ways to approach the problem.

- **Solutions to Problems** The student guide contains detailed solutions including the reasoning behind the solution to all the problems found in the *Genetics: From Genes to Genomes* textbook.

What additional things you can do to help yourself learn genetics?

- There is a great deal of new vocabulary and you may find it useful to make a list of the vocabulary that is new to you (along with the definitions). Try to weave together the new vocabulary in a narrative describing the main points of the chapter. You may have difficulty when you try to explain certain concepts. That is an indication that you don't really have as solid an understanding of that concept as you should have. The first problem of each chapter should help you evaluate your comfort with key vocabulary terms.

- Go to your professor with questions about a difficult concept. Work on describing where you have questions. Trying this sometimes leads you to discover the answer. Being able to identify what confuses you is a very valuable general skill.

- Become an active learner as you read new material. Stop every few pages to ask yourself: How would I summarize what I just read? How does it apply to what I have already learned? What do I not understand of this material?

- Make the learning process generally applicable – don't make problem-solving a drudgery. Solving genetic puzzles can be fun!

- You will find additional self-quizzes, study tips, vocabulary exercises, and web links to related topics by visiting the *Genetic: From Genes to Genomes* web site at www.mhhe.com/hartwell.

Good luck in the course!

Chapter 1   Mendel's Breakthrough: Patterns, Particles and Principles of Heredity

Synopsis

Chapter 1 covers the basic principles of inheritance as first described by Mendel. This chapter contains the basic concepts that form the foundation of inheritance (Laws of Segregation and Independent Assortment). You will see in chapter 3 how these laws relate to chromosome segregation during meiosis. Chapter 1 also contains most of the essential terminology used to describe inheritance. You should become very familiar with and fluent in the use of these terms because they will be used in increasingly sophisticated ways in subsequent chapters. A good way to assure that you have a solid grasp of the meanings of the new terms is to pretend you are describing each word or phenomenon to a friend or relative who is not a science major. Often giving an example of each term is useful. The first problem at the end of the chapter is also a useful gauge of how well you know these terms.

A few of the terms defined in this chapter are very critical yet often misunderstood. Learn to be precise about the way in which you use these terms:

genes and alleles of genes (A gene determines a trait; there are different forms or alleles of a gene. The <u>pea color gene</u> has two forms or alleles: <u>yellow</u> and <u>green alleles</u>.)

genotype and phenotype (Genotype is the genetic makeup of an organism (given as alleles) and phenotype is what the organism looks like.)

dominance and recessive (When one dominant allele is present, its phenotype will be expressed.)

homozygous and heterozygous (When both alleles of a gene are the same, we say the organism is homozygous for that gene; if the two alleles are different, it is heterozygous)

Be prepared to:

After reading the chapter and thinking about the concepts you should be able to:

- set up Punnett squares (determine gametes produced by parents and the potential offspring)
- recognize if a trait is dominant or recessive
- recognize the ratio of phenotypes from a cross between two individuals heterozygous for one gene (3:1 ratio)
- recognize the ratio of phenotypes from a cross between two individuals heterozygous for two genes (9:3:3:1 ratio)
- recognize the need for and be able to set up a test cross
- determine the genotypes and phenotypes of offspring and the ratio of each, starting with genotype of parents
- determine the genotype of the parents, starting with the phenotypes of offspring

- determine probabilities using the sum and product rule
- draw and interpret pedigrees

Problem solving tips:

- Don't let the problem overwhelm you. Take each sentence and rewrite the information given in a format that will be useful for you.
- Read the problem carefully and closely- wording is important and often critical for interpreting the information that has been given to you and what is being asked.
- Start by designating alleles and genotypes- it will help you sort out the information given to you in the problem.
- If you are describing gametes in a problem, remember there is only one allele of each gene per gamete.
- When describing genotypes of individuals, remember there are two alleles of each gene.
- Think about whether you need to look at genotypes or phenotypes
- Answer the question asked! Always go back to the original problem and make sure you have done this.
- Basic rules of probability (keep in mind the simple dice examples)
  Product rule: If two outcomes must occur together, the probability of one outcome AND the other occurring is the product of the two individual probabilities. (The final outcome is the result of two independent events.) So, the probability of getting a 4 on one die AND 4 on the second die is the product of the two individual probabilities
  Sum rule: If there is more than one way in which an outcome can be produced, the probability of either one OR the other occurring is the sum of the individual probabilities. (The outcomes are mutually exclusive in this case).

Social and ethical issues

There are no obvious, easy, "right" answers to many of the ethical issues. The viewpoints of each individual, the impact on individuals involved and society overall must be considered and weighed carefully. People will hold different, valid opinions about these issues, and it is important to hear viewpoints of others.

In discussing what can be done in the situations described in the Social and Ethical Issues , consider the following:

- Who are the players?
- Who is involved in making the decisions?
- Who will be impacted by the decision made or action taken?
- What values are important for each of these players?
- What are the options?

A course of action may have to be a compromise for individuals involved.

Solutions to Problems

1-1. a) 4 b) 3 c) 6 d) 7 e) 11 f) 13 g) 10 h) 2 i) 14 j) 9 k) 12 l) 8 m) 5 n) 1

1-2. Peas have several advantages for studying inheritance. Peas 1) have a rapid generation time 2) can self-fertilize or be artificially crossed by an experimenter 3) large numbers of offspring could be obtained 4) several traits with two easily distinguished, discrete forms existed 5) can be maintained as pure-breeding lines and 6) are easy to grow. In contrast, 1)the human generation time is long
2) there is no self-fertilization and manipulated crosses are not ethical 3) numbers of offspring per mating are not great 4) some human traits have two very distinct forms but many show a continuum of phenotypes 5) individuals that are homozygous for a trait (analogous to pure-breeding) exist but are not "maintained" 6) require lots of maintenance to "grow".

1-3. The dominant trait (short tail) is easier to eliminate from the population by selective breeding. You can recognize every animal that has inherited the allele, because only one dominant allele is needed to see the phenotype. Those mice that have inherited the dominant allele can be prevented from mating. The recessive mutation, on the other hand, can be passed unrecognized from generation to generation in heterozygous mice (carriers). The heterozygous mice do not express the phenotype, so cannot be distinguished from normal homozygous dominant mice. Only the homozygous recessive mice express the dilute phenotype and could be prevented from mating.

1-4. a) Two genetically different eggs could be produced by the *AabbCCDD* woman. Gametes from a woman of genotype *AabbCCDD* would always contain the *b*, *C*, and *D* forms of these genes because she is homozygous for these three genes. Her gametes could contain the either the *A* or *a* allele. The two types of gametes are *AbCD* and *abCD*.
b) Four genetically different eggs are possible. All gametes would contain *A* and *d* alleles but there are two possible *B* alleles *(B or b)* and two possible *C* alleles *(C or c)* the could end up in the gametes. $2 \times 2 = 4$.
c) Eight genetically different eggs are possible. Three of the genes are heterozygous so there are two forms of each of these genes that can be combined in gametes. $(2 \times 2 \times 2 = 8)$
d) Sixteen genetically different eggs are possible $(2 \times 2 \times 2 \times 2 = 16)$.

1-5. a) The alleles in the egg and sperm combine during fertilization to form a zygote. The zygote must have a genotype *aaBbCcDDEe*.
b) The gametes could be *aBCDE*, or *aBcDE*, or *aBCDe*, or *aBcDe*, or *abCDE*, or *abCDe*, *abcDE*, or *abcDe*

1-6. Designate alleles and genotypes. *D* = dominant dimple allele, *d* = nondimpled. Since the chin dimple phenotype is dominant, there are two genotypes, *DD* or *Dd* that result in the dimpled phenotype. The nondimpled individuals are homozygous recessive (*dd*).

a) The genotype of a nondimpled man is unambiguous (*dd*). His wife, who has the chin dimple, could be either *DD* or *Dd*. We are told her mother lacked the dimple (*dd*) so could only pass on the *d* allele to her daughter (the wife). The wife therefore has the *Dd* genotype. There is a 1/2 chance that a child would be *Dd* (and therefore have the chin dimple) and a 1/2 chance that a child would be *dd*. Overall, we would expect 1/2 of their children to have the chin dimple.

b) The nondimpled woman must be *dd*; her husband's genotype is ambiguous (*DD* or *Dd*). Because they have a nondimpled child (*dd*), the husband must also have a *d* allele to contribute to her offspring. He has genotype of *Dd*.

c) All eight dimpled children must have inherited the *D* allele from their father, since their mother (nondimpled, *dd*) could only contribute a *d* allele. The father could be either *DD* or *Dd* but it seems likely, given that large number of offspring that must have inherited the *D* allele from him, that he has the *DD* genotype with only the *D* allele to contribute. We cannot rule out the possibility that he is heterozygous (*Dd*) and by chance in each separate event the *D* allele was in the sperm that fertilized the egg but the probability is low that this would occur. (The probability of all children inheriting the *D* allele from an *Dd* parent is $(1/2)^8$ or 1/256.)

1-7. Because two phenotypes result from the mating of two cats of the same phenotype, the short-haired parent cats must have been heterozygous or hybrid. The phenotype expressed in the hybrids (the parent cats) is the dominant phenotype. Therefore short-hair is dominant to long-hair.

1-8. a) You are told that two affected individuals had one child with the piebald condition and one lacking the condition. This would not be possible if the affected individuals were homozygous. The affected parents must be heterozygous and the phenotype expressed in the heterozygous individual is the dominant trait. The piebald trait is therefore dominant.

b) The genotype of a person with piebald spotting could be either homozygous *(PP)* or heterozygous *(Pp)*. For the two affected individuals described here to have an unaffected child *(pp)*, they would have to have both been heterozygous *(Pp)*.

1-9. a) Look for the data that most clearly gives you a clue about the inheritance. The resulting offspring from the dry × dry individuals in the population are the most definitive and recognizable. Because there are only dry earwax offspring from this cross, we can conclude that all those dry parents were homozygous (pure-breeding). When a phenotype is recessive, all individuals that have that phenotype must be homozygous. Therefore dry is recessive; sticky is dominant.

b) A 3:1 ratio comes from crosses between two heterozygous individuals. For this trait, individuals with the *Ss* genotype would have the phenotype of sticky earwax so sticky × sticky would include matings between heterozygotes but could also include *SS* x *Ss*. The sticky × sticky matings studied in this human population would be a mix of matings between two heterozygotes (*Ss* × *Ss*), between two homozygotes (*SS* × *SS*) and between a homozygote and heterozygote (*SS* × *Ss*) because there is no way to distinguish the different types of crosses based on the phenotypes. The 3:1 ratio of the heterozygote cross is therefore obscured by being combined with results of two other crosses. A similar argument can be made for the 1:1 ratio. A 1:1 ratio comes from a heterozygous sticky (*Ss*) × homozygous dry (*ss*) but the sticky x dry matings here would include not only the *Ss* × *ss* but also homozygous sticky (*SS*) × homozygous dry (*ss*).

1-10. a) 1/6

b) 3/6 or 1/2. There are three possible even numbers (2,4, or 6). The probability of obtaining any one of the mutually exclusive (one die is thrown so only one of these possibilities can occur.) results is 1/6 + 1/6 +1/6. The sum rule is used here because of the mutually exclusive outcomes (also notice the or in the possible outcomes)

c) 2/6 or 1/3. A 6 or 3 is evenly divisible by 3. Therefore there is a 1/6 +1/6 or 2/6 = 1/3 chance of obtaining a number divisible by 3.

d) 1/36. When rolling a pair of dice, the resulting number on each die is independent of the other. Therefore we use the product rule to predict probability of the two independent events. 1/6 x 1/6 = 1/36

e) There are three possible even numbers you could get on one die and each occurs with the probability of 1/6. The overall probability of an even number on one die is 1/6+1/6+1/6 or 1/2. Similarly, the probability of getting an odd number is 1/6+1/6+1/6=1/2. The probability of both occurring is 1/2 × 1/2 = 1/4.

f) The probability of a 1 on one die is 1/6. The probability that a 1 will appear on the other die is 1/6. The probability of both occurring is 1/6 × 1/6 or 1/36. The same is true for the other 5 possible numbers on the dice. The probability of any one of these mutually exclusive situations occurring is 1/36 + 1/36 +1/36 + 1/36 + 1/36 + 1/36= 6/36 or 1/6.

g) The probability of getting two numbers both over four is the probability of getting a 5 or 6 (1/6 + 1/6) on one die and 5 or 6 on the other die (1/6 + 1/6). 2/6 × 2/6 = 4/36.

1-11. This problem asks for the probability of occurrence of two traits: sex of a child and galactosemia. The parents are heterozygous for galactosemia so there is a 1/4 chance that a child of theirs will be affected (homozygous recessive). That probability that a child is a girl is 1/2. The probability of an affected girl is therefore (1/2)(1/4) = 1/8.

a) The fraternal (non-identical) twins result from two independent fertilization events and therefore the probability that both will be girls with galactosemia is the product of their individual probabilities. 1/8 × 1/8 = 1/64.

b) If the twins are identical, one fertilization event gave rise to two individuals. The probability that both are girls with galactosemia is 1/8.

1-12. a) 2/16 or 1/8 probability of a child like AaBbCcDd. The child would have to inherit all the dominant alleles from parent 1. For each of these 4 genes there is a 1/2 probability of a dominant allele being in a gamete ($1/2 \times 1/2 \times 1/2 \times 1/2$ or 1/16). The recessive alleles would be contributed by parent 2.

To produce a child with a genotype like parent 2, all recessive alleles would have to be in the gamete from parent 1 (probability = $1/2 \times 1/2 \times 1/2 \times 1/2$ or 1/16) as well as the in the gamete from parent 2 (probability = 1). Because there are two possible outcomes, the probability that the progeny have the phenotype of <u>either</u> parent is the sum of the individual probabilities: 1/16 + 1/16 = 2/16.
b) One parent is homozygous for the dominant alleles of all four genes; the other is homozygous for the recessive alleles. They cannot have a child with the same genotype but can have one that has the same phenotype as the second parent (AABBCCDD). A heterozygous child (AaBbCcDd) will have the same phenotype. The probability of this occurring is 1 because all gametes from the first parent will have the recessive alleles and from the second parent will have the dominant allele.
c) For a child to phenotypically resemble either of the parents the genotype would have to be *A-B-C-D-*. The probability of getting at least one dominant allele of each gene is 3/4 for each gene. (Remember the 3/4:1/4 ratio of phenotypes for a monohybrid cross.) The probability of one dominant allele for each gene is (3/4)(3/4)(3/4)(3/4)=81/256.
d) All children will resemble the parents phenotypically (as well as genotypically).

1-13. A fly with the dominant phenotype (normal wings) can have either a homozygous (*WW*) or heterozygous (*Ww*) genotype. The genotype can be determined by performing a testcross; that is, crossing your fly with the dominant phenotype (but unknown genotype) to a fly with the recessive (short wing) phenotype (and therefore the homozygous recessive genotype,*ww*). If the fly with the dominant phenotype has the homozygous dominant genotype, the cross would be *WW* × *ww* and all the progeny in this case would be *Ww* and would have the dominant phenotype. If the fly being tested had a heterozygous genotype, the cross would in fact be *Ww* × *ww* . 1/2 of the progeny would be *Ww* and have the dominant phenotype, and 1/2 of the progeny would be *ww* and have the recessive phenotype. Using the results of the cross, you can deduce the genotype of the original fly.

1-14. The results of the crosses all fit very well the pattern of inheritance of a single gene, with the closed trait being recessive. The first cross, closed x open, yielding all open is a cross like Mendel did with his pure-breeding plants (although we don't know from this problem if the starting plants were pure-breeding). The results of the first cross suggest that open is dominant and self-fertilization of the F_1 resulting in a 3:1 ratio of open to closed supports this hypothesis. The cross of the F_1 x closed resulted in the 1:1 ratio seen for a testcross in which a heterozygote is crossed to a homozygous recessive. All the data is consistent with open being recessive.

1-15. No. When a cross between animals of different phenotypes results in offspring having each of two phenotypes, the cross was between heterozygous and homozygous recessive animals. But, this does not indicate whether red or black is the dominant phenotype. An animal with a recessive phenotype would have to be pure-breeding (homozygous genotype). If you mate several red horses to each other and also mate several black horses to each other, the crosses that always yield offspring with the parental phenotype must have been between homozygous recessives. For example, if all the black x black matings result in black offspring, black is recessive. The red × red crosses would then have resulted in both red and black offspring.

1-16. The F_1 must be heterozygous for all the genes because the parents were pure-breeding (homozygous). From a heterozygous cross, both dominant and recessive phenotypes can be seen. For these four traits, all combinations of dominant and recessive versions of each trait are possible. The possibilities can be determined using a forked-line method shown below.

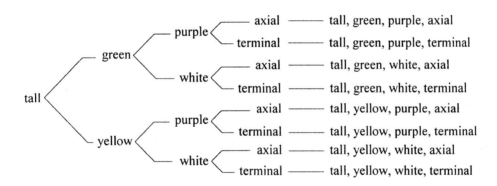

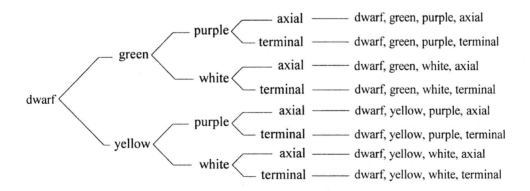

b) Designate the alleles: T = tall; t = short

G = green; g = yellow

P = purple p = white

The F$_1$ plant is heterozygous ($TtGgPpAa$.). The dwarf parent was pure breeding and therefore must be homozygous for every trait ($ttggppaa$) In this cross, the $TtGgPpAa$ plant could produce 16 different types of gametes in equal proportions (1/16 of each). To determine the genotypes, use the forked line method as in a). A gamete could contain either T or t; each of these different classes of gametes could contain either G or g, etc. (The frequency with which each gamete type is expected can be calculated using the product rule. There is a 1/2 probability of either allele and the probability for any particular combination is therefore 1/2 × 1/2 × 1/2 × 1/2 or 1/16).

Gametes from $TtGgPpAa$

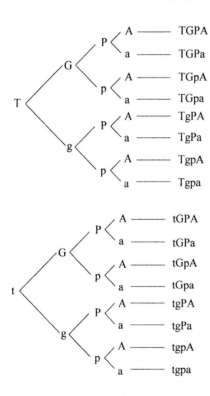

The only gametes from the $ttggppAA$ parent would be $tgpA$. The progeny expected from the cross are in the same ratios as the gametes: 1/16 of each type.

1-17. To solve this problem, look for characteristic phenotypic ratios of offspring. Consider not only the ratios of the two traits together but also look at the ratios for each separate trait. The first cross gives information on which alleles are dominant. The results are very close to a 9:3:3:1 ratio that would be seen in a dihybrid cross (heterozygote × heterozygote for two genes). The phenotypic class that contains the largest number of individuals in such a cross has the dominant phenotypes. Designate the alleles P = purple, p = white; S = spiny, s = smooth.

a) The first cross is *PpSs* x *PpSs* (explained above).

b) There are only two phenotypic classes (purple, spiny and purple, smooth). Notice that all progeny are purple. There was a 1:1 ratio of spiny to smooth pods. A 1:1 ratio arises from heterozygote × homozygous recessive *(Ss × ss)*. Because all progeny were purple, at least one parent plant must have been homozygous for the P allele. The cross was *PPSs × P⁻pss* or *P⁻Ss × PPss*.

c) Only two phenotypic classes are seen here with all progeny being spiny. The 1:1 ratio of purple to white indicates that the parents were *Pp × pp*. Because all progeny were spiny, at least one parent was homozygous for spiny *(SS)*. The genotypes of the parents are *PpSS × ppSS* or *PpS⁻ × ppSS* or *PpSS × ppS-*

d) The ratios of the four phenotypes do not fit the 9:3:3:1 ratio of a cross between two dihybrids. Looking at each trait individually, there are 89 +31 or 120 purple plants; 92 +27 or 119 white plants. The 1:1 ratio that results from crossing a heterozygote to a homozygous recessive. There are 89 + 92 or 181 spiny pods; 31 +27 = 58 smooth pods. This is close to a 3:1 ratio indicating that the spiny pod plants were heterozygous for the *S* gene. The genotypes of the parents were *ppSs × PpSs*.

e) All progeny have smooth pods so the parents were both homozygous recessive. Among these progeny is a 3:1 ratio of purple to white plants. The parents must have been heterozygous for the P gene. The genotypes of the parents are *Ppss × Ppss*.

f) There is a 3:1 ratio of spiny to smooth podded plants, indicative of a cross between heterozygotes *(Ss x Ss)*. All progeny plants were white so the parents must have been pure-breeding recessives. The genotypes of the parents are *ppSs × ppSs*.

1-18. In offspring (seeds), 3/4 are round and 1/4 are wrinkled. The parents must have been heterozygous (*Rr*) for the pea shape gene. All the offspring are yellow and therefore have the *Yy* or *YY* genotype. To get only these genotypes, the parent plants (phenotypically yellow) could have both been *YY* or one *YY* and one *Yy*.

1-19. a) Since only one phenotype was seen in the first cross, we can assume that the F_1 generation consists of hybrid (heterozygous) animals and the phenotype of these animals indicates the dominant allele. Rough and black are the dominant alleles. (*R* = rough, *r* = smooth; *B* = black, *b* = white) The findings of the F_2 generation support this conclusion.

b) An F_1 male is heterozygous for both genes with a genotype *RrBb*. The smooth white female with both recessive phenotypes must be homozygous recessive *(rrbb)*. When this male is mated with the smooth white female (*rrbb*) the gametes would be *RB* or *rb* or *rB* or *Rb* from the F_1 male and *rb* from the female. The four genotypes possible are *RrBb*, *Rrbb*, *rrBb*, and *rrbb* and correspond to the phenotypes rough, black; rough, white; smooth, black; smooth, white respectively. These four phenotypes would be seen in the offspring in equal proportions (1:1:1:1).

1-20. Three characters are being analyzed in this cross. We can usually tell which alleles are dominant from the phenotype of the heterozygotes. In this problem, we are not told the phenotype of the heterozygote. The dominant phenotypes can also be recognized from the most prevalent phenotypic class from crossing heterozygotes. Look at each trait individually. Consider height first. There are 272 + 92 + 88 + 35 or 487 tall plants : 93 + 31 + 29 +11 or 164 dwarf plants. This is close to a 3 :1 ratio expected from the heterozygous cross and indicates that tall is dominant. Now consider pod shape. 272 + 92 + 93 + 31 or 488 inflated pods : 88 + 35 + 29 + 11 or 163 flat pods. Inflated is dominant. Finally consider, flower color. 272 + 88 + 93 + 29 + 11 = 493 purple flowers : 92 + 35 + 31 + 11 = 169 white flowers. Purple is dominant.

1-21. a) The probability that a child will have the achoo syndrome (dominant) and non-trembling chin (recessive) is the product of their individual probabilities because inheritance of the two genes are separate events. The probability of two heterozygous individuals having a child with the dominant trait is 3/4; with the recessive trait is 1/4. 3/4 × 1/4 = 3/16.
b) 1/4 (probability of a non-achoo) × 1/4 (probability of a non-trembling chin) = 1/16

1-22. The parents of the F_1 plant were *YYrr* and *yyRR* plants. The F_1 has the genotype *YyRr*. When it is allowed to self-fertilize, the yellow round phenotype will appear with a 9/16 probability. Each pea results from a separate fertilization event. The probability that seven of these fertilizations to occur in one pod (giving 7 yellow round peas) would be 9/16 × 9 /16 × 9/16 × 9/16 × 9/16 × 9/16 × 9/16 or 4,782,969/268,435,456 or .018.

1-23. a) Recessive. Two unaffected individuals have an affected child and it was a consanguineous marriage that produced the affected child. Affected=*aa;* carrier=*Aa*
b) Dominant. The trait is seen in each generation; affected person has an affected parent; III-3 is unaffected even though both his parents are (this would not be possible for a recessive trait). Affected=*AA*; carrier- not applicable
c) Recessive. Two unaffected parents have an affected child. Affected=*aa;* carrier=*Aa*

1-24. Designate alleles: *P*-normal pigmentation allele; *p*-albino allele
An albino must have the homozygous recessive genotype, *pp.* The parents are normal in pigmentation and therefore could be *PP* or *Pp.* Because they had an albino child, they must both have the *Pp* genotype. The probability that their next child will be albino is completely independent of the first child's genotype. From a cross between two heterozygotes, three genotypes can result: *PP, Pp,* and *pp.* The probability of the *pp* genotype is 1/4.

1-25. It is easiest to work these problems if you draw the information into a pedigree.

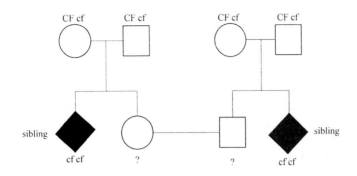

a) Each member of the married couple had a sibling affected. Therefore their parents must have been heterozygous carriers. Heterozygous parents can have affected children (homozygous), carriers of recessive alleles (heterozygous) or normal individuals (homozygous) with a probability of 1/4, 1/2, 1/4. Because the woman being counseled is unaffected, we know she does not have the homozygous recessive genotype. Of the remaining possibilities 2/4 of the 3/4 remaining are heterozygous. There is therefore a 2/3 chance that she is a carrier.

b) This couple can only have an affected child if both are carriers. The probability that both are carriers is 2/3 × 2/3 or 4/9. The probability that they will have an affected child if 1/4 × 4/9 (probability that both are carriers) = 4/36 or 1/9.

c) A child of this couple could be a carrier if both parents are heterozygous or if one is heterozygous and one is homozygous dominant. As was calculated above, the probability that both parents were heterozygous is 2/3 × 2/3. The probability that two heterozygous parents will have a heterozygous child is 1/2. The probability of being heterozygous and having a heterozygous child is therefore (2/3 × 2/3 × 1/2) or 4/18 or 2/9. The probability that one parent is heterozygous (2/3) and the other is homozygous (1/3) and that they have a heterozygous child is 2/3 × 1/3 × 1/2 or 1/9. The probability of either one occurring is the sum of these mutually exclusive events. 1/9 + 2/9 = 3/9.

1-26. a) The affected father could be either homozygous or heterozygous for the Huntington disease allele. The problem indicates that the disease if rare, therefore is if most likely that the father is heterozygous *(Hh)*. Because the disease is dominant, the son can be expected to develop symptoms if he inherits the *H* allele. There is a 1/2 chance that he inherited this allele from his father.

b) We don't know the genotype of the man in his twenties but he has a 1/2 chance of having inherited the Huntington allele and his children have a 1/2 chance of inheriting either of his alleles, so the chance they will inherit the disease allele and develop symptoms is 1/2 × 1/2 or 1/4.

1-27. Pairs of unaffected individuals (I-1 and I-2; II-3 and II-4) had affected children (II-1, III-1, III-2). The disease looks like it is recessive. Because there are two cases in which an individual outside of the bloodline must have been a carrier, the disease allele appears to be common in the population.

Chapter 2   Extensions to Mendel: Complexities in Relating Genotype to Phenotype

Synopsis

This chapter builds on the principles of segregation and independent assortment that you learned in chapter 1. An understanding of those basic principles will help you understand the more complex inheritance patterns in chapter 2. While the basic principles hold true, the expression of the phenotype is more complicated.

There are several examples in this chapter of single gene inheritance in which phenotypic ratios are different from the complete dominance examples in chapter 1:

- incomplete dominance
- codominance
- dominance series of multiple alleles
- lethal alleles

Also introduced in this chapter are examples in which two or more genes determine the phenotype. Crosses involving two genes that interact to produce a phenotype have a phenotypic ratio in the F_2 generation that is a modification of the genotypic ratio of 9:3:3:1. Multigene (>2 genes involved) inheritance leads to many more phenotypic classes. (There is more information on multigene inheritance in chapter 10).

Penetrance and expressivity are terms used to describe some cases of altered phenotypic expression. Penetrance describes the number of individuals with a mutant genotype who are affected and expressivity describes the extent to which individuals with a mutant genotype are affected.

Be prepared to:

After reading the chapter and thinking about the concepts you should be able to:

- distinguish single and two gene traits on the basis of inheritance patterns (look for a some variation of 9:3:3:1 in the F_2 to identify a dihybrid inheritance)
- know how to set up crosses to determine whether one or two genes are involved in expression of the phenotype
- based on offspring ratios, be able to propose ways in which two genes interact. (Do not merely memorize the altered ratios-think through what the combinations of alleles mean).
- recognize incomplete dominance and lethal alleles based on offspring ratios in the F_1 and F_2. These ratios are easily recognized, but you should also understand how to derive these ratios if you have forgotten them.

Problem solving tips

- Novel phenotypes arise when there is codominance or incomplete dominance. In these cases of single gene inheritance, the novel phenotype occurs in the F_1 generation. If two genes are involved in determining a trait, you may see a novel phenotype in the F_1 but an additional new phenotype may also appear in the F_2 generation (in contrast to single gene inheritance).

- It is critical that you understand the 9:3:3:1 genotypic ratio for a dihybrid cross. Know what each class represents (9 A-B-, 3A-bb, 3 aaB-, 1 aabb, where a dashed line (-) indicates either a dominant or recessive allele)

- Remember the product rule of probability and use it to determine proportions of genotypes or phenotypes.

- Make sure you are keeping genotypes and phenotypes straight when working the problems.

Solutions to Problems

2-1. a) 2 b) 6 c) 11 d) 8 e) 7 f) 9 g) 10 h) 3 i) 5 j) 4 k) 1

2-2. This problem involves incomplete dominance, recognized by the intermediate phenotype. Designate alleles: *R*=red; *r*= white. Genotypes: *RR*= red; *Rr*= pink; and *rr* = white.
a) The genotype of pink four o'clocks is *Rr*. *Rr x Rr* would result in a 1:2:1 ratio of red: pink: white. Note that the phenotypic ratio is the same as the genotypic ratio in incomplete dominance.
b) The cross of white x pink is *rr* x *Rr*, and yields *Rr* and *rr* plants. The ratio of phenotypes among the offspring will be 1:1 pink: white.
c) A red × red cross is *RR x RR* and will produce all red flowered plants (*RR*).
d) A red × pink cross is *RR x Rr* and produces 1:1 ratio of red (*RR*) and pink (*Rr*).
e) The white × white cross *(rr x rr)* produces all white flowered plants *(rr)*.
f) A red × white cross *(RR x rr)* produces all pink *(Rr)* flowered plants.

2-3. The ratio of offspring looks like 1:2:1. What cross gives these results? Crosses between organisms heterozygous for incompletely dominant genes produces a 1:2:1 ratio of phenotypes in the next generation.   The flower color gene therefore shows incomplete dominance. The yellow parents are heterozygous for flower color *(Yy)* and yield 1/4 *YY* progeny (red), 1/2 *Yy* progeny (yellow) and 1/4 *yy* progeny (white).

2-4. For each of these traits that show incomplete dominance we know the ratios of phenotypes in the F$_2$ generation (1:2:1). That means there will be 1/4 red: 2/4 purple: 1/4 white and 1/4 long: 2/4 oval : 1/4 round.   List all the possible phenotype combinations and use the product rule of probability to determine the ratios of phenotypes in the F$_2$.

Phenotype	Probability of phenotype
red, long	$1/4 \times 1/4 = 1/16$
red, oval	$1/4 \times 1/2 = 1/8$
red, round	$1/4 \times 1/4 = 1/16$
purple, long	$1/2 \times 1/4 = 1/8$
purple, oval	$1/2 \times 1/2 = 1/4$
purple, round	$1/2 \times 1/4 = 1/8$
white, long	$1/4 \times 1/4 = 1/16$
white, oval	$1/4 \times 1/2 = 1/8$
white, round	$1/4 \times 1/4 = 1/16$

2-5. When encountering a problem involving crosses of two traits, examine each trait independently. The total number of long plants is 301 + 612 + 295 = 1208/ 1600 total plants. There are 99 + 195 + 98 = 392 short plants. The ratio is very close to 3:1 long : short plants. Therefore the *L* (long) allele is completely dominant to *l* (short) allele. The second trait is flower color. There are three phenotypes in the F$_2$, which should alert you to the possibility of incomplete dominance. Check to see if the ratios fit those expected for incomplete dominance. There are 301 + 99 or 400 purple; 612 + 195 or 807 pinks; 295 + 98 = 393 white. The ratio of purple : pink : white is very close to 1:2:1, therefore the flower color trait shows incomplete dominance of two alleles.

2-6. a) Hb^{S}Hb^S

b) A child with the sickle cell phenotype must be homozygous and therefore inherited a mutant allele from each parent. Because the parent mentioned is normal, he/she must also have a normal allele, therefore the genotype of that parent is *Hb^{S}Hb^A*.

c) The number alleles possible in the children is dependent on the number of alleles in the parents. Each individual has two copies of the β-globin gene so could carry two different alleles. If each parent were heterozygous for different alleles, there are four possible alleles that could be found in the five children.

2-7. a) An O girl has the genotype *ii*. Each parent had to have an *i* allele to contribute. Parents could be *ii* (phenotype O) or *iI^A* (phenotype A) or *iI^B* (phenotype B)

b) A girl with blood type B could have an $I^B I^B$ or $I^B i$ genotype. The mother is type A and therefore could not have contributed an I^B allele (but must have contributed the *i* allele). The father must have contributed the I^B allele (and could be $I^B I^B$, $I^B i$ or $I^B I^A$).

2-8. a) father is B MN Rh pos b) father is O MN Rh pos c) father is A M Rh pos d) father is O M Rh neg

To approach this problem, look at the mother/child combination to determine what alleles the father must have contributed to that child's genotype. For example, for a) the male had to contribute I^B N and *Rh neg* alleles. Only the male fitting these requirements is the male with the phenotype B MN and Rh pos (and he must be $Rh^+ Rh^-$).

b) The male had to contribute *i*, *N* and *Rh neg* alleles. The father could be either the male O MN Rh pos or B MN Rh pos. However the latter is the only male fitting the requirements for the mother and child in a), therefore the father in b) must be O MN Rh pos (assuming one child per male).

c) The male had to contribute I^A *M* and *Rh neg* alleles. Only the male who is A M Rh pos fits these criteria.

d) The male had to contribute either I^B or *i*, *M* and *Rh neg*. Three males could have the alleles required: the O M Rh neg male, the O MN Rh pos male and the B MN Rh pos male. However only the O M Rh neg male remains unassigned to a mother child pair.

2-9. Designate the alleles: *L1* = marbled; *L2* = spotted; *L3* = dotted; *L4* = clear

For the marbled *(L1L1)* x spotted *(L2L2)* cross, the F_1 will be all marbled *(L1L2)*. For the dotted *(L3L3)* x clear *(L4L4)* cross, the F_1 will be all dotted *(L3L4)*.

a) To determine the number of phenotypes in the cross between an F_1 plant of each of these crosses *(L1L2 x L3L4)*, determine the genotypes possible. The *L1L2* plant can produce *L1* or *L2* gametes, the *L3L4* plant can produce *L3* or *L4* gametes in equal proportions. Four genotypes are possible: *L1L3, L1L4, L2L3, L2L4*. Progeny *L1L3* and *L1L4* will both be marbled because marbled is dominant to both the dotted and clear alleles. Progeny *L2L3* will have both spots and dots. Progeny *L3L4* will be spotted. There will be three different phenotypes: spotted and dotted; marbled; and spotted in a 1:2:1 ratio.

b) The F_1 from the first cross are marbled *(L1L2)* and dotted *(L3L4)*.

2-10. Designate two different alleles for plant 1 (You know from the incompatibility system that each plant is heterozygous).

<u>alleles</u>

plant 1 alleles   1 , 2

Looking at the results of crosses, plants 2, 3, and 5 must have one (or both) of the alleles from plant 1 since they are incompatible with plant 1 (produce no seeds) but you don't know which allele (1 or 2). Plant 4 has neither allele present in plant 1. Let's give plant 4 alleles 3 and 4. Looking at plant 2 crosses, plant 2 does not share alleles with 3, 4 or 5 so it must have the allele found in plant 1 that plant 5 does not contain. Designate allele 1 as the allele shared between plants 1 and plant 2 and designate allele 2 as the allele shared between plants 1 and 3 and 5. Plant 3 crosses show that it shares an allele with plant 5 (already designated allele 2) but it does not share an allele with plant 4. Call the second allele number 5. Plant 4 × 5 indicates that plants 4 and 5 share an allele, say allele 3. We still don't know a second allele for plant 2 but it is not any of those carried by plants 3, 4, or 5, so it must be another allele. There are six alleles total.

plant	genotype
1	1,2
2	1,6
3	2,5
4	3,4
5	2,3

2-11. The cross is between a A^yA Cc × and a A^yA cc mouse. When crosses involve two genes in which there is a modification of the ratios expected from simple dominance, consider each trait independently and use the product rule to determine the phenotypic ratios. Looking first at the A locus, we expect a 1:2:1 genotypic ratio of A^yA^y, A^yA, AA but A^y is a homozygous lethal allele, so we wouldn't see any of this class. That leaves a 2: 1 (or 2/3: 1/3) of A^yA : AA. The Cc x cc portion of the cross results in 1/2 Cc and 1/2 cc offspring. All cc mice will be albino regardless of A alleles. Each of these classes could have either the Cc or cc alleles a

$$2/3 \ A^yA \quad \times 1/2 \ Cc \quad = 2/6 \ A^yACc \ \text{(yellow)}$$
$$\times 1/2 \ cc \quad = 2/6 \ A^yAcc \ \text{(albino)}$$

$$1/3 \ AA \quad \times 1/2 \ Cc \quad = 1/6 \ AACc \ \text{(agouti)}$$
$$\times 1/2 \ cc \quad = 1/6 \ AAcc \ \text{(albino)}$$

So overall there will be 2/6 yellow, 3/6 albino, and 1/6 agouti progeny.

2-12. a) The 2:1 phenotypic ratio together with the statement that montezumas are never true-breeding suggests that there is a lethal allele of this gene. When there is an allele that is lethal when homozygous, crossing two heterozygous individuals results in a 1:2:1 genotypic ratio, but one of those classes is lethal. The result is the 2:1 phenotypic ratio as seen in this cross. The montezuma parents were therefore heterozygous, *Mm*.

b) Designate the alleles: M = montezuma, m = greenish

F = normal fin, f = ruffled

The cross is *MmFF* x *mmff*. The progeny will be 1/2 *MmFf* (montezuma, normal fin) and 1/2 *mmFf* (greenish, normal fin).

c) The cross is :*MmFf* x *Mm Ff*

The progeny will be:

2/3 montezuma	X	3/4 normal fin	= 6/12 montezuma, normal fin
	X	1/4 ruffled fin	= 2/12 montezuma, ruffled fin
1/3 green	X	3/4 normal fin	= 3/12 green, normal fin
	X	1/4 ruffled fin	= 1/12 green, ruffled fin

2-13. a) We know some information about the genotypes of the these two mice from the phenotype. The rest we have to deduce based on the progeny of the mice. The yellow mouse must have an A^y allele, but we don't know the second allele of the A gene. A^y is epistatic to the B gene so we don't know what alleles the yellow mouse has at the B gene. Since the mouse does show color we know it is not *cc* (albino), but must have at least one C allele. The brown agouti mouse has at least one A allele, but the other could be *a* or *A*; must be *bb* at the B gene; and again, since there is color, is not *cc* but *C-*. The mating between these two can be represented as A^y? ?? C? × A? bb C? Because one pup was albino (*cc*), the parents must have been heterozygous *Cc*. A brown mouse (*bb*) indicates that both parents had to be able to contribute a *b* allele, so we now know the first mouse must have had at least one *b* allele. And since the brown mouse is not agouti, the parents must have each carried an *a* allele. Because a black agouti mouse was born, we know that the first mouse must have also had a *B* allele. The complete genotypes of the mice are:

$A^y a Bb Cc$ × *Aa bb Cc*

b) Six different coat color phenotypes are possible: albino, yellow, brown agouti, black agouti, brown, black. Think about each gene individually, then the effect of the other genes in combination with that phenotype: *Cc* or *cc* leads to a phenotype with or without color (albino). The genotypic classes for the A gene are $A^y A$, $A^y a$, *Aa* and *aa*. The first two would have the same yellow phenotype. Since yellow is epistatic to B, we know this is a class that will be seen regardless of the genotype of the B gene. This is one phenotypic class: yellow. *Aa* is agouti and the type of agouti in coloration depends on the B gene. For the B gene, the offspring could be *Bb* or *bb* (black or brown).

This means that together with the agouti (*A*) allele, the coat color phenotype will be brown or black agouti. With the *aa* genotype, there is no yellow on the hair, so the combinations of *B* alleles (*Bb* or *bb*) would result in brown or black coats.

2-14. a) The ratio of individuals with four different phenotypes in the F$_2$ is close to 9:3:3:1. Whenever you see this ratio, you should think of a cross involving two genes. In this example, the two genes could be referred to as *A* and *B*. Because the results in the F$_2$ show that the walnut phenotype is most abundant (the 9 of the 9:3:3:1 ratio), walnut must be the phenotype that results from dominant alleles of both genes (*A-B-*). Single combs are the lowest in number and is therefore represent the homozygous recessive genotype (*aabb*). If there is a dominant allele of one gene (designate it *A*), the phenotype is rose; if there is a dominant allele of the other gene (*B*) the phenotype is pea.

b) A homozygous rose hen would be *Aabb*. When crossed to a homozygous pea combed hen, *aaBB*, the result is the doubly heterozygous F$_1$ *AaBb* that would have a walnut comb. In the F$_2$ you again expect the 9:3:3:1 ratio of walnut, rose, pea, and single combs.

2-15. With four phenotypes appearing in the F$_2$ generation, there must be two genes involved in determining coat color. Another clue to determining genotypes is that crossing a liver colored horse to either one of the original parents resulted in the parent's phenotype. The liver horse's alleles do not seem to be contributing to the phenotype which is the case when the alleles are recessive *(aabb)*. Because the F$_1$ bay colored animals produce four phenotypic classes, they are heterozygous (*AaBb*). This is consistent with the result that their parents were homozygous for each of the two genes involved. The black mare was *AAbb* and the chestnut stallion was *aaBB*, the liver horses were *aabb*, the bay horses were *AaBb*.

2-16. a) Because unaffected individuals had affected children, the trait appears to be recessive. From affected individual II-1, you know the mutant allele is present in this generation. The trait was passed on through II-2 who was a carrier. All children of affected individuals III-2, III-3 are affected as predicted for a recessive trait. However, generation V seems inconsistent with recessive inheritance of a single gene. It is consistent with two different genes involved in hearing with a defect in either gene leading to deafness. The two family lines shown contain mutations in the separate genes and the mutations are recessive.

b) Individuals in V would be heterozygous *AaBb* having inherited from their parents (*aaBB* and *AAbb*) a dominant and recessive allele of each gene. The one copy of each gene product is sufficient for normal function (This is an example of complementation of two genes).

2-17. The 9:7 ratio comes from the 9:3:3:1 ratio with the 3 (*A-bb*) and 3 (*aaB-*) and 1 (*aabb*) classes combined into one phenotypic class of yellow (recessive genotype for either gene leads to yellow color). The heterozygous F$_1$ (*AaBb*) when test crossed to homozygous (*aabb*) would produce the following genotypes: *AaBb, aaBb, Aabb,* and *aabb* in equal proportions (¼ of each). Only the *AaBb* would have green fruit so ¼ would be green; ¾ would have yellow fruit.

2-18. a) No, a single gene cannot account for this result. While the 1:1 ratio seems like a testcross ratio (homozygous recessive × heterozygous), the fact that the phenotypes of the offspring are not the same two phenotypes seen in the parents argues against this being a testcross.

b) The appearance of four phenotypes suggests involvement of two genes.

c) The 3:1 ratio suggests that two alleles of one gene differentiate between the wild-type and scattered patterns.

d) The true-breeding wild-type fish are homozygous and the scattered fish have to be homozygous recessive according to the ratio seen in c) so the cross is:

> *bb* × *BB*

The F$_1$ will be heterozygous *(Bb)* and the F$_2$ will be 1/4 *BB*, 1/2 *Bb* and 1/4 *bb* (3/4 wild-type and 1/4 scattered).

e) The inability to obtain a true-breeding nude stock suggests that the nude fish are heterozygous and that the fish homozygous for the nude allele (designate that *A*) die. The nude × nude is *Aa* × *Aa* and the *AA* homozygotes die, leaving a 2:1 ratio of the *Aa* and *aa* genotypes, corresponding to nude and scattered phenotypes.

f) Going back to the linear cross, the fact that there are four phenotypes led us to propose two genes were involved and the ratio looks like an altered 9:3:3:1 ratio. Variations of the 9:3:3:1 ratio are produced when heterozygotes are crossed, so the linear parents are doubly heterozygous. How could the 6:3:2:1 ratio arise? Loss of three genotypes usually found in the 9 category and one genotype in the phenotype 3 category could produce this ratio. From the results in e) we know the *AA* genotype is lethal and this fits with the 6:3:2:1 ratio seen. The phenotypes and corresponding genotypes are:

linear	*AaB-*
nude	*Aabb*
wild-type	*aaB-*
scattered	*aabb*

2-19. Dominance relationships are between alleles of the same gene. Only one gene is involved. In contrast, epistasis involves two genes. The alleles at one gene affect the expression of a second gene.

2-20. When new phenotypes appear in F$_2$ and the numbers are some variation of 9:3:3:1 ratio, it means two genes are involved in determining the trait.

A-B-	9	$\Rightarrow$	13 : 3
A-bb	3	$\Rightarrow$	
aa B-	3		
aabb	1	$\Rightarrow$	

The 9 *A-B-*, 3 *A-bb*, and 1 *aabb* could have the same phenotype, so the ratio would be 13:3. This would occur if there is dominant epistasis with the dominant *A* allele hiding expression of the *B* genotype. Plants with an *A* allele have the same phenotype as those plants with the *aabb* genotype.

2-21. The $I^A I^B Ss$ parent can produce four different types of gametes: $I^A S$, $I^A s$, $I^B S$, $I^b s$. The $I^A I^A Ss$ parent can produce two kinds of gametes: $I^A S$ or $I^A s$. The expected offspring and apparent phenotypes would be:

1/4 $I^A S$	$\times$	1/2 $I^A S$ = 1/8 $I^A I^A SS$	A
	$\times$	1/2 $I^A s$ = 1/8 $I^A I^A Ss$	A
1/4 $I^A s$	$\times$	1/2 $I^A S$ = 1/8 $I^A I^A Ss$	A
	$\times$	1/2 $I^A s$ = 1/8 $I^A I^a ss$	O
1/4 $I^B S$	$\times$	1/2 $I^A S$ = 1/8 $I^A I^B SS$	AB
	$\times$	1/2 $I^a s$ = 1/8 $I^A I^B Ss$	AB
1/4 $I^B s$	$\times$	1/2 $I^A S$ = 1/8 $I^A I^B Ss$	AB
	$\times$	1/4 $I^A s$ = 1/8 $I^A I^B ss$	O

All individuals with the *ss* genotype look like type O - combining other classes - 1/4 would appear to have O type blood, 3/8 have A, 3/8 have AB.

2-22. *Aa Bb* parents will have offspring in the genotypic ratios 9 *A-B-*; 3*A-bb*; 3*aaB-*; 1*aabb*. But, since the defect in enzyme is only seen if both genes are defective, only the *aabb* genotype will result in abnormal progeny. Combining the ratios of individuals in the other classes, the phenotypic ratio is 15:1.

2-23. a)

	I-1	I-2	I-3	I-4	II-1	II-2	II-3	III-1	III-2
blood type	AB	A	B	AB	O	O	AB	A	O
genotypes	$I^A I^B$	$I^A ?$	$I^B ?$	$I^A I^B$	*ii*	$I^A ?$	$I^A I^B$	$I^A ?$	??
	Hh	*Hh*				*hh*	*Hh*		*hh*

When you try to assign blood type genotypes in the pedigree, you find there are inconsistencies between expectations and what could be inherited from a parent. For example, I-1 (AB) × I-2 (A) could not have an O child (II-2). The epistatic *h* allele (Bombay phenotype) could explain these inconsistencies. If II-2 has an O phenotype (does not produce A or B antigens) because she is homozygous for the *h* allele, her parents must have been heterozygous *Hh*. The Bombay phenotype would again explain that seeming inconsistency of two O individuals (II-1 and II-2) having an A child. II-2 could have received an *A* allele from one of her parents and passed this on to III-1 together with one *h* allele. Parent II-1 would have to contribute the *H* allele so that the *A* allele would be expressed. Individuals II-2 and II-3 could not have had an $I^O I^O$ child since II-3 had $I^A I^B$ genotype, but III-3 has the O phenotype. II-3 must be *Hh* and III is *hh*. (A ? in the table above indicates that there is more than one allele possible.)

2-24. The difference between pleiotropic mutations and traits determined by several genes would be seen if crosses were done using pure-breeding plants, then crossing the F_1 progeny (selfing). In the F_2 generation, there would be several different combinations of the petal color, markings and stem position if several genes were involved. If all traits were due to the allele present at one gene, the three phenotypes (e.g., yellow petals, brown markings and erect stems) would always be inherited together.

2-25. For each gene there is a 3/4 probability of getting at least one dominant allele from crossing flies heterozygous as three loci.

$3/4A \times 3/4B \times 3/4C = 27/64$ wildtype

$1-(27/64) = 37/64$ mutant

2-26. a) The 9:7 ratio in the F_2 comes from the necessity of a dominant allele of each of two genes to get the red color. The White × White → red and 9:7 ratio in the F_2 indicates that the mutations causing the white phenotype are in separate genes. So, 1 and 2 are in separate genes as are 1 and 3; 2 and 3. Therefore, three genes are involved.

b) Designate the genes 1, 2, and 3 and the alleles + (dominant red allele) or - (recessive white allele) for each gene:

Gene	alleles
1	$1^+, 1^-$
2	$2^+, 2^-$
3	$3^+, 3^-$

The genotypes of the white mutant strains are:

Strain	Genotype
White-1	$1^- 1^- 2^+ 2^+ 3^+ 3^+$
White-2	$1^+ 1^+ 2^- 2^- 3^+ 3^+$
White-3	$1^+ 1^+ 2^+ 2^+ 3^- 3^-$

c) Genotype of cross   F₁ Progeny   F₂ progeny:

$1^-1^-2^+2^+$ × $1^+1^+2^-2^-$ → $1^+1^-2^+2^-$ → $1^+\text{-}2^+\text{-}$ 9/16   red

$1^+\text{-}2^-2^-$ 3/16   white

$1^-1^-2^+$ -   3/16   white

$1^-1^-2^-2^-$ 1/16   white

(Gene 3 is wild-type in both strains, so you don't need to include this gene in your cross.) The second, third, and fourth classes in the F₂ progeny are homozygous recessive for either gene 1 or 2 and have the same phenotype (white).

2-27. You are asked to consider the number of affected children in a subset of all the children (the live-born). First determine the genotypes of the children.

Genotypes of progeny:

1/2 $D1^+/D1^m$ × 1/4 $D2^+/D2^+$ × 1/2 $D3^+/D3^m$ =1/16 normal

× 1/2 $D3^m/D3^m$ =1/16 deaf $D3^m/D3^m$

× 1/2 $D2^+/D2^m$ × 1/2 $D3^+/D3^m$ =1/8 or 2/16 normal

× 1/2 $D3^m/D3^m$= 2/16 deaf $D3^m/D3^m$

× 1/4 $D2^m/D2^m$ × 1/2 $D3^+/D3^m$ = 1/16 deaf $D2^m/D2^m$

× 1/2 $D3^m/D3^m$ = 1/16 deaf

double mutant (see below)

1/2 $D1^m/D1^m$ × 1/4 $D2^+/D2^+$ × 1/2 $D3^+/D3^m$ = 1/16 deaf $D1^m/D1^m$

× 1/2 $D3^m/D3^m$ = 1/16 deaf

double mutant

× 1/2 $D2^+/D2^m$ × 1/2 $D3^+/D3^m$ = 2/16 deaf $D1^m/D1^m$

× 1/2 $D3^m/D3^m$ = 2/16 deaf

double mutant

× 1/4 $D2^m/D2^m$ × 1/2 $D3^+/D3^m$ = 1/16 deaf

double mutant

× 1/2 $D3^m/D3^m$ = 1/16 deaf

triple mutant

Look for the single, double and triple mutant combinations:

7/16 are singly mutant: 3/16 $D1^m/D1^m$

1/16 $D2^m/D2^m$

3/16 $D3^m/D3^m$

5/16 are doubly mutant: 1/16 $D2^m D2^m D3^m D3^m$

3/16 $D1^m D1^m D3^m D3^m$

1/16 $D1^m D1^m D2^m D2^m$

Of the 5/16 with double mutations, 1/4 are lethal (25%), and 3/4 are deaf. So 5/16 × 3/4 =15/64 are live deaf children with double mutations.

1/16 are triply mutant but 75% of these die (1/16 × 3/4= 3/64) and only 1/16× 1/4 = 1/64 live and are deaf.

Converting all the numbers to 64ths, there are 3/16 or 12/64 normal; 7/16 or 28/64+15/64 +1/64=44/64 deaf; and 5/64+3/64=8/64 die. So there will be 12 normal: 44 deaf live born children, so the likelihood of a deaf child is 44/56.

2-28. a) The pattern looks like recessive since unaffected individuals have affected progeny, but given that the trait is rare you wouldn't expect two heterozygotes to marry by chance in as many cases as seen here. The alternative explanation is that the trait is dominant but not 100% penetrant.

b) II-3, III-6 would be non-expressers in the Smiths, II-6 in the Jeffersons.

c) When common, the recessive inheritance is most likely mode of inheritance.

d) Now II-3, II-4; II-6, III-7 in the Smiths family (both parents have to he heterozygous to have an affected child) and II-6 and II-7 in the Jeffersons family.

Chapter 3   The Chromosome Theory of Inheritance

Synopsis

Chapter 3 is very critical for understanding basic genetics because it connects chromosome behavior during meiosis with the laws of heredity described by Mendel. While you may have learned mitosis and meiosis in your basic biology class, now is the time to make sure you understand these processes in the context of inheritance. The physical basis for inheritance is chromosome segregation during meiosis. You should have an increased understanding of the importance of meiosis for genetic diversity through both independent assortment and recombination.

Make sure you understand the following two statements from this chapter:

"in the first meiotic division the centromeres do not divide as they do in mitosis"

"Sister chromatids contain the same genes but may carry different combinations of alleles"

The experiments that showed the correlation between chromosome behavior and inheritance using X-linked genes in *Drosophila* are described in this chapter. X-linked traits have characteristic inheritance patterns recognized in pedigrees or in results of reciprocal crosses.

Be prepared to:

After reading the chapter and thinking about the concepts you should be able to:

- draw chromosome alignments during metaphase of mitosis, meiosis I, meiosis II
- describe how chromosome behavior explains the laws of segregation and independent assortment
- identify and distinguish between homologs and sister chromatids
- identify sex-linked inheritance patterns
- determine genotypes in sex-linked pedigrees and probabilities of specific genotypes and phenotypes
- determine if non-disjunction occurred in meiosis I or II and the parent in which it occurred based on the genotype of a child

Problem solving tips:

- Keep clear the distinction between sister chromatids (identical, replicated copies of a chromosome) and homologs (chromosomes carrying the same genes but different alleles).
- When you see different numbers of male and female progeny for a particular phenotype, you might suspect that the gene is located on a sex chromosome.
- Two features that tip you off about X-linked inheritance (especially in pedigrees) are criss-cross inheritance (inheritance of a characteristic from mother to son and father to daughter) and a greater number of males than females affected.
- Remember that sons receive their X chromosome from their mother and have to pass on their X chromosome to their daughter.

Solutions to Problems

3-1. a) 5 b) 7 c) 11 d) 10 e) 12 f) 8 g) 9 h) 1 I) 6 j) 4 k) 3 l) 2

3-2. a) Two daughter cells are produced by mitosis with 14 chromosomes each. The chromosome number is maintained after mitotic divisions.
b) Meiosis would produce four cells, each with 7 chromosomes (one half the number of chromosomes of the starting cell).

3-3. a) 23 chromosomes come from the father.
b) Each somatic cell had 44 autosomes and 2 sex chromosomes.
c) A human ovum (female gamete) contains 23 chromosomes.
d) One sex chromosome is present in a human ovum.

3-4. The diploid sporophyte contains 7 pairs of chromosomes, with one chromosome in each pair from the male gamete, the other from the female gamete. The probability that any one specific chromosome that was contributed from the father ends up in a particular haploid spore is 1/2. The probability that all chromosomes were originally from the male is $(1/2)^7$.

3-5. a) The ivory-eyed queen must have a homozygous genotype (*bb*) because she has a recessive phenotype. When crossed to a brown-eyed drone (*B*), the resulting offspring would all be *Bb* or brown females. The males that development parthenogenetically from the *bb* females would be ivory-eyed (*b*).
b) A female from the first cross would be *Bb*. Crossed to a *B* drone, all the progeny would be female with brown eyes (with genotypes *BB* and *Bb*). Parthenogenetic males from the *Bb* male would be ivory (*b*) and brown-eyed (*B*).

3-6. a) G_1, S, G_2 and M.
b). G_1, S and G_2 are all part of interphase.
c) G_1 is the time of major cell growth that precedes chromosome replication. Chromosome replication occurs during S phase. G2 is another phase of cell growth after chromosome replication during which the cell synthesizes many proteins needed for mitosis.

3-7. a) iii b) i c) iv d) ii e) v

3-8. a) 48 chromosomes containing 2 chromatids each = 96 chromatids.
b) 48 chromosomes × 2 chromatids each = 96 chromosomes
c) 24 chromosomes (the chromosome number is halved by the end of meiosis I) × 2 chromatids = 48 chromatids.

d) 48 chromosomes (unreplicated in G_1) = 48 chromatids

e) 48 chromosomes × 2 chromatids = 96

f) 48 chromosomes (unreplicated in G_1 preceding meiosis) = 48 chromatids

g) 48 chromosomes × 2 chromatids each = 96 chromatids

3-9. a) mitosis, meiosis I, II

b) mitosis, meiosis I

c) mitosis

d) meiosis I

e) meiosis I

f) none

g) meiosis I

h) meiosis II, mitosis

i) mitosis, meiosis I

3-10. Remember that the problem states that all cells are from the same organism. This influences the designation of mitosis, meiosis I and II for some of the figures. The *n* number is 3 chromosomes.

a) meiosis I

b) mitosis (not meiosis II! because there are 6 chromosomes or 2n in this cell)

c) meiosis II

d) mitosis

e) meiosis II

3-11. When trisomy 21 cells undergo meiosis, there would be an extra unpaired chromosome 21. This chromosome would segregate randomly into one of the products of meiosis I. When that cell goes through meiosis II it would have an extra chromosome 21 and if one of those two meiotic products is the ovum, a trisomic child would be produced after fertilization. There is a 2/4 chance that an ovum will contain the extra chromosome 21.

3-12. The genetic reshuffling that occurs during meiosis based on independent alignment of maternal and paternal chromosomes and recombination between homologs could lead to collections of alleles of genes that help an organism survive.

3-13. a) 400 spermatozoa are produced from 100 primary spermatocytes.

b) 100 spermatozoa are produced from 100 spermatids.

c) 100 ova are formed from 100 primary oocytes (remember that although each primary oocyte will produce three of four meiotic products (depending on whether the first polar body undergoes meiosis II), only one will become an egg (ovum)).

d) No ova are produced from polar bodies.

3-14. The primary oocyte contains a duplicated set of the diploid number of chromosomes. During meiosis I, the homologous chromosomes segregate into two separate cells, so the chromosome carrying the *A* allele will segregate into one cell, the chromosome carrying the *a* allele will segregate into the other cell. One of the cells becomes the secondary oocyte (containing more of the cytoplasm) and the other becomes the polar body. The genotype of the dermoid cyst that develops from a secondary oocyte could be either *AA* or *aa*.

3-15. a) The eggs (ZW) would give rise to only females (ZW).
b) Cells resulting from meiosis would be either Z or W; then upon duplication would become ZZ or WW. ZZ cells develop into males. Since WW is lethal, only males are produced by this mechanism.
c) After eggs have gone through meiosis I, they will contain either Z or W chromosomes composed of sister chromatids that separate to become the chromosomes. Again only males (ZZ) are produced since WW cells are inviable.
d) An egg produced by complete meiosis could be either Z or W and could fuse with either a Z or W polar body to result in ZZ (males) and ZW (females) or the inviable WW. To calculate frequencies, the probability of a Z egg is ½ and the probability of a W egg is ½. If it is a Z egg, then there is a 1/3 probability that one of the 3 polar bodies will be Z, and a ⅔ probability that it will be W (because you already used up a Z in the egg). If it is a W egg, then it is ⅓ W and ⅔ Z for the polar bodies.
Frequency of obtaining genotype:
ZZ: ½ × ⅔   ZW: ½ × ⅔   WW: ½ × ⅓   WZ: ½ × ⅔
ZZ: 1/6   WW: 1/6   WZ: 4/6
Because WW is inviable, among the surviving chicks, the ratio would be :
Males (ZZ) = 1/5   Females (WZ) = 4/5

3-16. The results of the second cross show crisscross inheritance (brown females × yellow males resulting in brown sons and yellow daughters). This is characteristic of sex-linked traits, so check to see if this fits with the data. In birds males are ZZ, females are ZW.
Because the cross between true-breeding birds gave brown males and females, brown is dominant.
Z^B=brown allele on Z chromosome
Z^b=yellow allele on Z chromosome
The first cross was $Z^B Z^B$ (brown male) × $Z^b W$ (yellow female) and produced $Z^B W$ (brown females) and $Z^B Z^b$ (brown males).
The second cross was $Z^B W$ (brown females) × $Z^b Z^b$ (yellow males) produced $Z^B Z^b$ (brown males) and $Z^b W$ (yellow females).

3-17. Look at each trait individually to determine if genes are X-linked and what allele is dominant. What would you predict if a gene was not sex-linked? The number of males and number of females with the same phenotype should be equivalent.

vestigial: 16 yellow males

                    15 brown males

                    31 brown females

normal: 48 yellow males

                    49 wildtype males

                    97 wildtype females

Because the numbers of males and females with vestigial wings are equivalent, the gene is located on an autosome. (The same is true for normal winged flies.) The ratio of normal winged flies to vestigial winged flies is close to 3:1 so the normal allele is dominant.

Looking at the body color trait,

yellow: 16 +48 + 64 males

                           no females

brown (wild-type):   15 + 49 = 64 males

                           31 + 97 = 128 females

The numbers of males and females of each phenotype are not equivalent so the gene is X-linked. The results can be explained if the trait is X-linked recessive. The absence of yellow females suggests that the yellow allele is recessive and contributed by the mother to the sons. The wild-type F_1 female flies must have been $X^B X^b$ and the wild-type F_1 males were $X^B Y$.

3-18. The red-eyed males are $X^{W^+}Y$. Nondisjunction in meiosis I in males results in a cell that contain both chromosomes and one lacking the sex chromosomes. After fertilization with an egg (X^W) from the white-eyed female, $X^{W^+}X^W Y$ flies (red-eyed female) and $X^W O$ (white-eyed male) flies would be produced respectively. If nondisjunction occurred in meiosis II, sperm would be produced that had $X^{W^+}X^{W^+}$ or YY (and those lacking a sex chromosome – null gametes). After fertilization, the zygotes would be $X^{W^+}X^{W^+}X^W$ (these flies die) or $X^W YY$ (white-eyed males). (Notice that the flies that live after these nondisjunction events have the same phenotypes as the normal flies (resulting from fertilization of gametes in which nondisjunction had not occurred) and therefore would be indistinguishable from the normal flies.)

3-19. a) In birds, females are the heterogametic sex, having a ZW composition. Z^B represents the Z chromosome carrying the dominant *B* (barred) allele. Crossing a barred hen ($Z^B W$) to a non-barred rooster (ZZ) would yield non-barred female chicks (ZW) and barred male chicks ($Z^B Z$).
b) Crossing an F_1 rooster ($Z^B Z$) to one of his sisters (ZW) would yield $Z^B W$ (barred) and ZW (non-barred) females and $Z^B Z$ and ZZ (barred and non-barred) males.

3-20. a) Remember that this problem stated that the pedigrees (i-iv) show examples of each of the four modes of inheritance. Pedigree i) represents an autosomal recessive trait because two unaffected individuals have affected children. It is not X-linked because both sexes are affected in the second generation. Pedigree ii) could represent either autosomal or X-linked inheritance but because i) must represent autosomal recessive inheritance, this pedigree represents X-linked recessive inheritance. Pedigree iii) appears to represent a dominant trait because there are affected individuals in both generations. The trait is autosomal because both sexes are affected in the second generation. (It could also be an autosomal recessive if the trait was common and I-2 was a carrier, but we already used the autosomal recessive trait for pedigree i). Pedigree iv) represents an X-linked dominant trait as characterized by the transmission from affected father to all daughters.

b) i) The parents must have been heterozygous carriers for this autosomal recessive trait. There is therefore a 1/4 chance that the child will be *aa* and have the trait.

ii) The woman in generation I, who is a carrier for this X-linked trait, has a 1/2 probability that she will pass on the X chromosome with the mutation to sons where it is expressed immediately or to daughters where it will not be expressed. (The unaffected father contributes a normal X chromosome.) There is a 1/2 chance that the mother (II-5) of the child in question carries the mutation. If the child is a son, there is a (1/2)(1/2) probability that the child is affected. If the child is female, there is no chance that it will be affected. (The father is unaffected and therefore cannot contribute an X carrying the mutant allele.)

iii) The expression of an autosomal dominant phenotype requires only one mutant allele. There is a 1/2 chance that the mother (II-5) will pass on the mutant allele to a child.

iv) The father (I-1) passes on the X chromosome carrying the mutation to all his daughters but none of his sons. II-5 therefore does not carry the mutation and cannot pass it on. The probability that the child will be affected is 0.

3-21. a) Albinism in this pedigree is caused by a recessive allele because the phenotype is not seen in each generation and two unaffected individuals have an affected child.

b) The trait is autosomal because if it were X-linked an affected female would have the mutant allele on both chromosomes and would have to pass the mutation on to sons. This does not happen with individuals I-40 or II-9.

c) *aa*

d) *Aa*

e) *Aa*

f) *Aa*

g) *Aa*

h) *Aa*

3-22. a) The boy received an X^B from his father but also a Y chromosome. The sperm therefore contained $X^B Y$ and nondisjunction must have occurred in meiosis I to result in both the X and Y segregating into the same daughter cell.

b) The son could have received both the X^A and X^B chromosomes from mother or the X^B from the mother and the X^A and Y chromosomes from the father. It is not possible to determine in which parent the nondisjunction occurred.

c) The son has two X^A chromosomes and a Y. He could only have received two X^A's from his mother and the nondisjunction could have occurred in either meiotic division.

3-23. Color-blindness is an X-linked recessive condition. The genotypes of the males can all be determined because the phenotype is dependent on the genotype of the one X chromosome. We know that individual II-2 and III-3 must be $X^{cb}Y$ because they are affected. Look at the parents of II-2. The mother must have contributed the X^{cb} to her son but she is unaffected, so her genotype is $X^{CB}X^{cb}$. The father (I-2) is normal and therefore $X^{CB}Y$. II-1 can be either $X^{CB}X^{CB}$ or $X^{CB}X^{cb}$. II-3 must be a carrier ($X^{CB}X^{cb}$) since she had an affected son. II-4 is $X^{CB}Y$; III-2 is either $X^{CB}X^{CB}$ or $X^{CB}X^{cb}$; III-4 is an unaffected male and therefore must be $X^{CB}Y$.

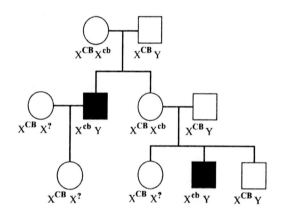

3-24. a) Draw the relationships into a pedigree.

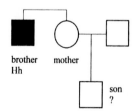

The mother of the woman and her affected brother must have been heterozygous for the *d* allele. The woman in the problem therefore has a 1/2 chance that she has one X^d chromosome. If her child is a son there is a 1/2 chance that he will receive the X^d chromosome from her. The overall probability is (1/2)(1/2)=1/4 that the woman will have an affected son.

b)

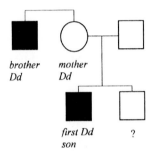

brother   mother
Dd   Dd

first Dd   ?
son

We now know that the woman is in fact a carrier since she had an affected son. Therefore the probability is 1/2 that she will pass on the X^d chromosome to her son and he will be affected.

c)

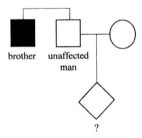

brother   unaffected
           man

?

The mother of the affected and unaffected man was heterozygous and passed on the X^d chromosome to the son who was affected but not to the unaffected son. There is no chance that the unaffected man will pass the disease allele to his children.

3-25. a) The father would have to have contributed *i* or I^A, Rh^+, *M* or *N*, $Xg^{(a-)}$ alleles. Only alleged father #3 fits these criteria.
b) If the daughter was XO (Turner), she could have inherited the one X from either mother or father. If the X came from her mother , that relieves the constraint on the allele her father has on the X

chromosome. The father could have the $Xg^{(a+)}$ genotype and therefore alleged father #1 or #3 fit the criteria for paternity.

3-26. Remember that white tigers are shown by unshaded symbols. Test each possibility by assigning genotypes based on the mode of inheritance.

a) No. Y linked genes are only expressed in males. Since there are white females, the trait could not be Y-linked.

b) Yes. White males would have to have white daughters but not white sons. The outcomes of all matings are consistent with a dominant X-linked allele.

c) Yes. The information in the pedigree is consistent with dominant autosomal inheritance.

d) No. Kesari would have to be $X^w X^w$ and Tony would be $X^W Y$. Their son would have to be $X^w y$ and would be white colored, but Bim is not white.

e) Yes.

3-27. This is a challenging experimental question. To search for mutations causing nondisjunction in females, you could take potential mutant females that are homozygous for y ($X^y X^y$) and mate them with $X^y/y^+ \cdot Y$ males. If there is no nondisjunction in the females, you would find yellow-bodied females (X^y/X^y) and brown-bodied males ($X^y/y^+ Y$) progeny. Nondisjunction during meiosis in the female would yield XX or nullo-X eggs. XX eggs fertilized with a $y^+ Y$-bearing sperm would be brown bodied $X^y/X^y/y^+ Y$ females. Nullo-X eggs fertilized by X-bearing sperm would be X^y/O yellow-bodied sterile males. (The other animals -XXX or OY- made from eggs resulting from nondisjunction would be dead.) Thus, X^y/X^y females yielding elevated frequencies of brown-bodied female and yellow-bodied male progeny from this cross are good candidates to harbor mutations affecting meiotic chromosome segregation.

Chapter 4   Linkage, Recombination and Mapping of Genes on Chromosomes

Synopsis

This chapter is devoted to a very important topic: linkage of genes. The concept of linkage is the basis for genetic mapping. Because genes on the same chromosome are physically connected or linked, the fate of alleles of those genes during meiosis is a partitioning into gametes in the same arrangement as found in the parents UNLESS there is recombination between them. The frequency with which recombination occurs is dependent on distance between the genetic loci. The greater the distance between two genes, the higher the probability that recombination will occur between them. This relationship between recombination frequency and distance is used to create genetic maps.

Tetrad analysis is done in certain yeast in which the meiotic products are kept together in a sac (ascus) so the results of a single meiosis are displayed in each tetrad (containing four spores). Working tetrad analysis problems will strengthen your understanding of chromosome segregation during meiosis.

Be prepared to:

After reading the chapter and thinking about the concepts you should be able to:

- determine if genes are linked or not based on the frequency of different types of gametes or progeny
- given the genotypes of parents and information on linkage of genes, list the gametes that can be produced (with and without recombination)
- determine the distance between linked genes based on the % recombination that occurred during meiosis
- determine map order of three genes
- determine if there is crossover interference in a three point cross
- determine whether results are statistically consistent with expected results using the chi square test
- identify different types of tetrads (parental ditype - PD, non-parental ditype - NPD, tetratype-T)
- determine linkage between genes given either random spore results or numbers of different types of asci
- identify centromere linked genes and determine distance from gene to centromere

Problem solving tips:

- numbers of recombinant gametes out of the total gametes gives you the % recombination which is translated into map distance (map units)

Three point crosses

- In a three point cross, designate the different gametes or offspring as noncrossover (parental), single crossover or double crosssover. Then determine the order of genes and the distance between each of the genes.
- The noncrossover classes will be represented by the greatest numbers of offspring
- The double crossover classes will be represented by the smallest numbers of offspring. (Sometimes one or both double crossover classes are missing.)
- You can set up all three possible arrangements of genes starting with parental combinations of alleles, do a double crossover for each arrangement and the order that gives you allele combinations equivalent to the double crossover class observed (smallest number of offspring) is the correct order.
- Alternatively, you can look at the double crossover class and see which of the three genes is now present with two alleles that both came from the other parent. (Some students find this method more intuitive and quicker than the previous method.) That gene will be the one found in the middle.

Tetrad analysis

- If the # PD tetrads is about equal to the # NPD tetrads, the genes are unlinked. (Independent assortment dictates that these two classes of asci will be about equal when genes are unlinked.)
- If the # PD tetrads is much greater than the # NPD tetrads, the genes are linked.
- Distance between linked genes is determined by calculating recombination frequency using the equation: RF= (NPD + 1)/2T/ total # tetrads × 100.
- Tetratype tetrads arise from single crossovers between two linked genes and from double crossovers involving three strands of DNA.
- Crossovers between a gene and the centromere mean that the two alleles are separated at the second meiotic division (second division segregation) instead of during the first meotic division. Centromere distance can be measured in yeasts with ordered asci, such as *Neurospora* and *Ascobolus*.

Solutions to Problems

4-1. a) 8 b) 4 c) 1 d) 11 e)2 f) 5 g) 6 h) 3 i) 10 j) 7 k) 9

4-2. Our null hypothesis is that there is independent assortment of genes yielding a phenotypic ratio of 9:3:3:1. There were 556 plants total, so calculate the expected numbers by multiplying $9/16 \times 556$; $3/16 \times 556$; $1/16 \times 556$.

	Expected	Observed
Round Yellow	313	315
Round Green	104	108
Wrinkled Yellow	104	101
Wrinkled Green	35	32

Using the chi square equation:

$$\frac{(315-313)^2}{313} \ + \ \frac{(108-104)^2}{104} \ + \ \frac{(101-104)^2}{104} \ + \ \frac{(32-35)^2}{35}$$

$$.01 \quad + \quad .15 \quad + \quad .09 \quad + \quad .26 \quad = 0.51$$

The number of classes is 4, so the degrees of freedom is 4–1 or 3. Using the table in the text, the probability of having obtained this level of deviation by chance alone is between .9 and .95 (90-95 %). The data are consistent with the null hypothesis and we therefore conclude that Mendel's data and hypothesis were a good fit.

4-3. a) To determine the genotype of the F_1 snake, use some of the information about the parent snakes. The parent snakes were orange (*O-bb*) and black (*ooB-*). Because the F_1 were all brown, we know that the orange snake could not have contributed an *o* allele (or there would have been some black snakes). The orange snake must be *OObb* and the black snake could not have contributed a *b* allele or there would have been some orange snakes. The black parent must be *ooBB*. We know that the F_1 must be *OoBb*.

b) The F_1 were heterozygous for both genes (*OoBb*). If the two loci assort independently, we expect a 9:3:3:1 ratio of brown: orange: black: albino from crossing the heterozygous F_1 snakes. The total number of progeny is 160. 9/16 or 90 of these progeny were expected to be brown; 3/16 or 30 were expected to be orange; 3/16 or 30 black; 1/16 or 10 albino.

c)   Expected   Observed
        90   100
        30   25
        30   22
        10   13

$$\frac{(100-90)^2}{90} \quad + \quad \frac{(25-30)^2}{30} \quad + \quad \frac{(22-30)^2}{30} \quad + \quad \frac{(13-10)^2}{10}$$

1.11   +   0.83   +   2.13   + $0.9 = 4.97$

There are three degrees of freedom (4 classes − 1) and the p value is between 0.5 and 0.1. The observed values do not differ significantly from the expected.

d) There is a 10-50% probability that these results would have been obtained by chance if the null hypothesis were true.

4-4. a) 1/4 of the F_2 mice would be expected to be dancers if the trait were determined by a recessive allele of one gene (*aa*).

b) If two genes were involved and the genotype had to be *aabb* for the trait to be expressed, 1/16 of the mice would be expected to be dancers.

c) Calculate the chi square values for each situation.

Null hypothesis: Dancing is caused by being homozygous for the recessive allele of one gene. (1/4 of the F_2 mice should be dancers). Calculating the expected numbers, $1/4 \times 50$ mice or 13 should have been dancers, 37 should have been nondancers.

	Expected	Observed
nondancers	37	42
dancers	13	8

$$\frac{(42-37)^2}{37} \quad + \quad \frac{(8-13)^2}{13}$$

.68   +   1.92   =   2.60

With one degree of freedom, the p value is between 0.5 and 0.1, so this null hypothesis is still a good fit with the data.

The alternative null hypothesis: Dancing is caused by being homozygous for the recessive alleles of two genes (*aabb*). (1/16 of the F$_2$ mice should be dancers).

	expected	observed
nondancers	47	42
dancers	3	8

$$\frac{(42-47)^2}{47} \quad + \quad \frac{(8-3)^2}{3}$$

$$.53 \quad + \quad 8.33 \quad = \quad 8.86$$

With one degree of freedom, the p value is < 0.005, so the null hypothesis is not a good fit. The one gene hypothesis is a better fit with the data.

4-5. a) First establish the genotype of the F$_1$ individuals. *AABB* was crossed to *aabb*, so the F$_1$ are *AaBb*, which can be represented as *AB/ab* to indicate the allelic combinations on the chromosomes (with *AB* on one of the homologs and *ab* on the other homolog). The gametes that can be produced are parental types: *AB, ab;* and recombinant types *Ab* and *aB*. Because the genes are 40 cM apart, the recombinants will make up 40% of the gametes. The *Ab* and *aB* gametes will each represent 20% of the gametes. The parental gametes together make up 60% of the gametes so each (*AB* and *ab*) will be found 30% of the time.

b) If the original (parental) cross was *AAbb* x *aaBB*, the allele combinations on each chromosome would be *Ab/aB*. Parental gametes in this case are *Ab* and *aB* and would each be found at 30%. The recombinant gametes *AB* and *ab* would be each be found 20% of the time.

4-6. a) The F$_1$ individuals have the genotype of *CD/cd*. They are crossed to the homozygous recessive parent and you were given the genotypes of the resulting progeny. Because the gamete from the homozygous recessive is always *cd*, we can effectively ignore a *c* and *d* allele from the F$_2$ progeny to make it easier to identify the classes of individuals. The two classes of individuals with the greatest numbers represent cases where parental type gametes were involved in the fertilization (*CD* or *cd* combining with the *cd* gamete from the homozygous recessive parent). The other two types of progeny result from fertilization of recombinant gametes (*Cd* or *cD* combining with the *cd* gamete from the homozygous recessive parent). The numbers of recombinants divided by the total number of offspring gives the map distance: (98 + 102)/(903 + 897+ 98 + 102) = 200/2000=10 % or 10 map units.

b) The F$_1$ from a *CCdd* x *ccDD* cross would be *Cd/cD*.   Using the recombination frequency determined from the other cross, recombinant gametes would make up 10%. The gametes produced by this individual would be 45% *Cd*, 45% *cD*, 5% *CD*, 5% *cd*. After fertilization, there would be 45% *Ccdd*, 45% *ccDd*, 5% *CcDd*, 5% *ccdd*.

4-7. To determine the probability that a child will have a particular genotype, we have to look at the gametes that can be produced by the parents. The man can only produce *ab* gametes, but the woman, with a genotype of *AB/ab*, can produce 40% *AB*, 40% *ab*, 10% *Ab*, 10% *aB*. Recombinant classes together total 20%. The probability of a child with the *Ab/ab* genotype is therefore 10%.

4-8. a) Designate the alleles: *D* = Huntington allele,   *d* = normal allele.
　　　　　　　　　　　　　　　　B = brachydactyly,   *b* = normal fingers

John's father is *bbDd*; his mother is *Bbdd*.

b) We know the John is *Bb* because he has brachydactyly. There is a 50% chance that he inherited the *D* allele but is not expressing the Huntington phenotype yet, or the *d* allele. His genotype could be *BbDd* or *Bbdd*. (The chance that John does not have the *D* allele is, in fact, higher than 50% since he has survived to 50 without showing symptoms, but we are solving the problem without considering this complication.)

c) The probability that the child will express Huntington is the product of the probability that John has the *D* allele (1/2) and the probability that he will pass it on to his child (1/2). The probability that the child will have Huntington and express the disease by age 50 is (1/2) (1/2)(2/3 penetrance) and the probability that the child will have and express brachydactyly is (1/2)(9/10). Together the probability of expressing both brachydactyly and Huntington phenotype is (1/2)(1/2)(2/3)(1/2)(9/10)=.07.

d) If the two loci are linked, the alleles on each of John's chromosomes will be either *Bd/bD* or *Bd/bd*. If it is the former, *BD* gametes could only be produced by recombination 10% of the time (representing one half of the 20% recombinant type gametes possible).   Since there is a 1/2 chance the John's genotype is *Bd/bD*, 1/2 × 1/10 = 1/20 represents the probability that John's child will have inherited both the Huntington and brachydactyly alleles. The probability of expressing both is (1/2)(1/10)(2/3)(9/10)=.03.

4-9. a) First determine the genotypes of the parents.

Designating alleles: *A* = normal pigmentation

　　　　　　　　　　　　　　a = albino allele

　　　　　　　　　　　　　　Hb^A = normal globin

　　　　　　　　　　　　　　Hb^S = sickle allele

The father has the genotype *aHb^A/aHb^A*and the mother has the genotype *AHb^S/AHb^S* . Because both traits are rare in the population, we assume that the man and woman are homozygous for the wild type allele of the gene dictating their normal traits (that is they are not carriers). The son has the genotype aHb^A/AHb^S. Given that the genes are separated by 1 map unit, recombinant gametes (2 types) should represent 1% of the gametes. His gametes would consist of:

49.5% aHb^A

49.5% AHb^S

0.5% aHb^S

0.5% AHb^A

b) The father has the genotype AHb^A/AHb^A and the mother has aHb^S/aHb^S. The daughter has the genotype of aHb^S/AHb^A and will produce

49.5% aHb^S

49.5% AHb^A

0.5% aHb^A

0.5% AHb^S

c) Sickle cell anemia results from an Hb^S/Hb^S genotype and albinism is due to aa genotype. From the male, there is a 0.5% chance of an aHb^S gamete and from the female, 49.5% chance. The probability is $.495 \times .005 = .002$

4-10. a) To determine if genes are linked, first predict the results of the cross if the genes are unlinked. In this case, a plant with blue, smooth kernels (A-W-) is crossed to a plant with yellow, wrinkled kernels ($aaww$). Since there are four classes of progeny in the F_1, the parent with blue, smooth kernels must be heterozygous for both genes ($AaWw$). From this cross, we would predict equal numbers of all four phenotypes in the F_1 if the genes were unlinked. Since the numbers are very skewed, with the smaller classes representing recombinant offspring, the genes are linked.
169 + 186 / (1447 + 169 + 186 + 1510) = 355/3312 or 10.7 % are recombinant, meaning the genes are separated by 10.7 map units.

b) The genotype of the blue smooth parent was $AaWw$. The arrangement of alleles in the parent is determined by looking at the phenotypes of the largest classes of progeny. Since blue, smooth and yellow wrinkled are found in the highest proportion, AW must be on one chromosome and aw on the other.

c) The blue, wrinkled F_1 has a genotype of Aw/aw and can produce Aw or aw gametes in equal proportions. Notice that recombination between these chromosomes yields the same two combinations of alleles (Aw and aw) as without recombination, so the recombinants are indistinguishable from parentals and each type of gamete is expected 50% of the time. The yellow smooth F_1 plant has a genotype of aW/aw. Again since recombination leads to the same genotypes, the frequency of each type is 50%. Four types of offspring are expected in equal proportions: Aw/aW, Aw/aw, aw/aW, aw/aw = blue, smooth; blue, wrinkled; yellow, smooth; yellow, wrinkled.

4-11. a) The homozygous brown rabbits (*CCbb*) were crossed to albinos with a genotype of *ccBB*. The F$_1$ were therefore *CcBb*. If the genes were unlinked, what progeny would you predict from the cross to *bbcc* rabbits? The F$_1$ should be able to produce *CB, cb, Cb* and *cB* gametes in equal proportions and after mating to animals that produce only *cb* gametes produce four genotypic classes of offspring: *CcBb, ccBb, Ccbb,* and *ccBB* rabbits representing the phenotypes black, albino, brown, and albino respectively. The ratios would be 1/4 black, 1/2 albino, 1/4 brown.
b) If *c* and *b* are linked, we would expect the classes *Cb* and *cB* to predominate in gametes. (*cb* and *CB* are recombinant gametes and are therefore present at lower levels.) These are represented in the F$_2$ by the *Ccbb* (brown) and *ccBb* (albino) classes. Since we cannot distinguish between the albinos resulting from fertilization of recombinant gametes (*cb*), and those resulting from parental gametes (*cB*), we have to use the proportion of the *CB* recombinant class (represented by black progeny) 34/200 = 17% and assume that the other class of recombinants *(cb)* is the same level (Since the crossing-over is a reciprocal exchange, this assumption is acceptable). The percent recombinants is estimated at 34% and the genes are 34 map units apart.

4-12. Two genes are involved in this cross but the frequencies of offspring do not look like frequencies expected for two independently assorting genes. The genes appear to be linked. Designate the alleles: *Cn* = wild type, *cn*= cinnabar; *Rd*= wild type, *rd*= reduced.
The homozygous wild-type female, *Cn Rd/Cn Rd* mated with the reduced cinnabar male *cn rd/ cn rd* and produces F$_1$ offspring with *Cn Rd/cn rd* genotype. What genotypes and phenotypes would be expected in the F$_2$ generation? Recombination occurs in females, not in males, so we know the males could only produce *Cn Rd* or *cn rd* gametes. Crossingover would generate two additional classes in the females (*Cn rd* and *cn Rd*).

female gamete	male gamete	progeny genotype	progeny phenotype
$cn^+ rd^+$	$cn^+ rd^+$	*Cn Rd/ Cn Rd*	wild type
	cn rd	$cn^+ rd^+/ cn rd$	wild type
cn rd	$cn^+ rd^+$	$cn rd / cn^+ rd^+$	wild type
	cn rd	*cn rd / cn rd*	cinnabar, reduced
$cn^+ rd$	$cn^+ rd^+$	$cn^+ rd / cn^+ rd^+$	wild type
	cn rd	$cn^+ rd / cn rd$	reduced
$cn rd^+$	$cn^+ rd^+$	$cn rd^+ / cn^+ rd^+$	wild type
	cn rd	$cn rd^+ / cn rd$	cinnabar

The reduced and cinnabar classes represent the results of recombination. Because reciprocal classes of progeny resulting from recombinant gametes have a wild-type phenotype (indistinguishable from other wild types) we can only assume that the reciprocal class of gametes was present in the same proportion. The numbers of recombinants = 7+7+9+9=32/400 total flies or 8% recombination. The genes are separated by 8 map units.

4-13. a) A, B, and C are alleles of the marker " gene", *D* and *d* are alleles of the disease gene. The marker and the disease locus are linked.   In the father, the *D* (disease allele) is on the same chromosome as the A form of the marker while the B form of the marker is on the same chromosome with the *d* allele. Recombination between the marker and the disease locus would result in the A form being on the same chromosome as the *d* allele and the B form on the same chromosome as the *D* allele.

a) If the *D* locus and marker are 0 m.u. apart, there is no recombination between them (they appear to be the same gene, in fact) so the probability that the child (who inherited the B form from the father) carries the *D* allele is 0.

b) If the distance between the disease locus and marker is 1 map unit, there is 1% recombination between the locus and marker. Because we are only considering cases where the child has the *B* marker, there is a 1% chance the child will have inherited the disease allele *D* instead of the parental *d* allele linked to *B*.

c) 5%.

d) 10%.

e) 50% .

4-14. a) Because you see only one of the phenotypes for each trait the parents must have been homozygous for all three genes and the F_1 plants are heterozygous. The phenotype of the heterozygotes indicates the dominant alleles: white, tall, normal-sized flowers.

b) Designate the alleles:

W = white,		*w* = red
P = normal-sized flowers,		*p* = peloria
T = tall,		*t* = dwarf

The genotype of the parents in the original cross was *WWPPTT* x *wwpptt*.

c) The double crossover classes (with the smallest number of progeny) are the dwarf class and the peloria, white class. These could only have arisen from a double crossover if the order of genes was *W-P-T* (*WPT/wpt*). Try the other orders to convince yourself. The dwarf, peloria class and the white class are parental classes (contain the largest numbers of progeny).   The dwarf, peloria, white class and the wild-type class arose from a single crossover between the *W* and *P* genes. Totaling the *W-P* recombinants classes: 56+48 + 6+ 5 (double crossovers also have a crossover in this region)=115/543 total = 21%.

The dwarf, white class and the peloria class arose from a single crossover between *P* and *T* genes. The total of single and double crossovers between *P* and *T* is 51+43+6+5=105/543 total = 19%. The map is:

W   21 m.u.   *P*   19 m.u.   *T*

d) The expected percentage of double crossovers is the product of the percentage of single recombination in each interval: (.19)(.21)=0.04. The observed frequency is 11/543 or .02. There is interference.

e) coefficient of coincidence = .02/0.04=0.5

4-15. (We assume no interference occurs in this cross). Map units indicate the percent recombination between two genes. Double crossovers will occur during meiosis in the parent $a^+b^+c^+$ and ab^+c female flies to generate a^+bc^+ and ab^+c gametes with a 0.2 x 0.1 = .02 or 2% frequency. There will be 10 flies of each of these classes. Crossing-over between a and b generates $a^+\ b\ c$ and $a\ b^+c^+$, and will occur 20% of the time, but since we already accounted for 2% of those crossovers between a and b in the progeny resulting from a double crossover event, there will only be 18%/2 for each of the two single crossovers between a and b or 90 of each type. Crossing-over between b and c generates a^+b^+c and abc^+ and occurs 10% of the time - 2% already accounted for in the double crossover class. There will be 8%/2 of each class or 40 flies. That leaves 720 flies in the parental classes or 360 $a^+b^+c^+$ and 360 abc.

4-16. To determine the distance between genes, total the classes of recombinants for each pair of genes. The parents are $\alpha + +$ and afg. Recombinants between the mating locus (a or α) and the f gene are 6 $a + g$, 6 $\alpha f+$, 1 $a + +$, 1 αfg for a total of 14 /101. Recombinants between the mating locus and the g gene are 13 $\alpha + g$, 14 $af+$, 1 $a + +$, 1 αfg for a total of 29/101. Recombinants between the f and g loci are 13 $\alpha + g$, 14 $af+$, 6 $a + g$, and 6 $\alpha f+$ for a total of 39/101. f and g loci must be the outer genes of the three with the mating locus between them. The distances are:

f_____14_____mating_____29_____g
                    (a or α)

4-17. a) Looking at the data, you should immediately become suspicious because there are four sets of phenotypic classes that have about equal numbers of individuals. Looking at the four classes having the largest numbers of individuals (black, scute, echinus, crossveinless; scute, echinus, crossveinless; wild type; black), they fall into pairs in which the only difference is the black vs normal allele. This can be explained by independent assortment of the black locus in relation to the other genes. Looking at the rest of the phenotypic classes, pairs of classes can be explained again by black being unlinked to the other genes. The other genes are linked, judging from the fact that they can be grouped into single crossover and double crossover classes (based on the numbers of individuals). The female must have been heterozygous for all the genes to generate the classes seen and because one of the parental classes is scute, echinus, crossveinless and the other is wildtype, the female's genotype was $sc\ ec\ cv/ + + +$ and $b/+$ on another chromosome.

b) To map the distances between the three linked genes, the paired classes (with black or wild-type allele on the other chromosome) should be combined. For example, the 653 black, scute, echinus, crossveinless and 650 scute echinus, crossveinless should be considered as one class. The double crossover class is scute, crossveinless; echinus. This could have arisen only if the original order was

sc   ec   cv

_____

+   +   +

The phenotypic classes (black), scute; (black), echinus, crossveinless result from single crossovers between *sc* and *ec*. The distance between these genes is the total of single crossovers in this region plus double crossovers: 71+73+73+74+1+1+1+1=295/3288= 0.089 = 9.0 map units.

The phenotypic classes (black) scute, echinus, and (black) crossveinless result from single crossovers between *ec* and *cv*. The total number of crossovers between these two genes is 87+84+86+83+1+1+1+1=344/3288(total)= 0.1046 = 10.5 map units.

c) Interference would result in a different number of double crossovers than predicted by the distance between genes.   Predicted crossovers =0.09 × 0.105=0.009

Observed frequency = 4/3288=0.001

There is interference.   0.001/0.009 = 0.11 $I = 1- C = 1- 0.11 = 0.89$

4-18. a) The appearance of two new phenotypes (red and white) indicate that flower color shows incomplete dominance. The pink flowered plants are *Pp*, red are *PP* and white are *pp*.

b) The expected ratio of red: pink : white would be 1:2:1. Calculating for the 650 plants, this equals 162.5 red, 325 pink, and 162.5 white.

c) Inheritance of anther color is analyzed by totaling the numbers of plants with black and with tan anthers independent of other phenotypes. Tan: 78+26+39+13+5+2 = 163/ 650. Black: 44 + 15 + 204 + 68 + 117 + 39 = 487/650. The ratio is about 3: 1 black : tan. Therefore black is dominant to tan. The same analysis for the stem length indicates that 487 plants had long stems and 163 had short stems. Again the ratio is about 3:1 and long is dominant to short stems.

d) Designate the alleles:

P = red   *p* = white

B =black   *b* =tan

L =long   *l* =short

Because the phenotypic ratios for each gene independently are characteristic of heterozygous crosses, the genotype of the original plant is *PpBbLl*.

e) If the stem length and anther color genes assort independently, the 9:3:3:1 phenotypic ratios should be seen in the progeny. Totaling all the progeny in each of the classes:

long   tan   122
short   tan   41
long   black   365
short   black   122

The ratio is close to a 9:3:3:1 ratio, so we can assume these genes are unlinked.

The expected ratios of progeny for flower color (incomplete dominance, 1:2:1 ratio) and stem length (complete dominance, 3:1 ratio) if the two genes were unlinked can be calculated using ratios predicted for each gene individually and the product rule.

$3/4$ long × $1/2$ pink = $3/8$ or $6/16$ long pink

$3/4$ long × $1/4$ red = $3/16$ long red

$3/4$ long × $1/4$ white = $3/16$ long white

$1/4$ short × $1/2$ pink = $1/8$ or $2/16$ short pink

$1/4$ short × $1/4$ red = $1/16$ short red

$1/4$ short × $1/4$ white = $1/16$ short white

A 6:3:3:2:1:1 ratio would be seen. The observed numbers are:

pink	long	243
red	long	122
white	long	122
pink	short	81
red	short	41
white	short	41

The ratio is close to that predicted; so the genes are unlinked.

The same analysis is done for flower color and anther color. The observed numbers are

red	tan	104
red	black	59
pink	tan	52
pink	black	272
white	tan	7
white	black	156

Because these numbers do not fit a 6:3:3:2:1:1 ratio, we can conclude that flower color and anther color are linked genes.

4-19. From the criss-cross inheritance of the recessive alleles from mother to sons, you can tell that all these genes are located on the X chromosome. Designate alleles:

dwp^+ = normal dwp =dwarp

rmp^+ = normal rmp =rumpled

pld^+ = normal pld =pallid

rv^+ = normal rv =raven

Cross 1 was *dwp rmp/dwp rmp* × *pld rv/Y* ⇒ *dwp rmp/pld rv* wild-type females and *dwp rmp/ Y* dwarp, rumpled males. (Only the recessive alleles are indicated in these genotypes.)

Cross 2 was *pld rv/ pld rv* × *dwp rmp/ Y* ⇒ *pld vr/ dwp rmp* normal females and *pld rv / Y* pallid, raven males.

Females from cross 1 were mated to recessive males: *dwp rmp/ pld rv × dwp rmp pld rv/Y.*
Notice that there are fewer classes of progeny than you would expect for four genes. With two alleles of each gene, you would expect 2×2×2×2 or 16 classes of progeny but only 8 are seen.

You never see classes that represent a recombination between *pallid* and *dwarp* loci (e.g., *pallid* and *dwarp* mutations in the same fly). This suggests that the two genes are so close together that there is essentially no recombination between the loci. (If a much larger number of progeny were examined, you might observe recombinants.) By considering *dwp* and *pld* as having the same location, the problem becomes a three-point cross between *dwp/pld, rv,* and *rmp.* The smallest classes (pallid; and dwarp raven rumpled) represent the double crossover class. This requires that raven, *rv,* is the middle marker:

dwp/pld^{+}	*rv*$^{+}$	*rmp*
dwp^{+}/*pld*	*rv*	*rmp*$^{+}$

The pallid raven rumpled and dwarp classes (47 + 48) represent single crossovers between the *rv* and *rmp* genes, while the pallid rumpled and dwarp raven classes (23 + 22) represent single crossovers between the *pld/dwp* and *rv* genes. Dwarp raven rumpled and pallid are the result of a DCO. Distance between A and R is 47 + 48 + 2 + 3 = 100/1000 = 10 map units. Distance between *pld/dwp* and *rv* = 23 + 22 + 3 + 2 = 50/1000 = 5 map units.

4-20. In cross 1, the number of parental and nonparental ditypes are almost equal and the number of tetratypes is very low so we can conclude that there is independent assortment -the *ad* and mating type loci are not linked. In cross 2 the number of parental ditypes is much greater that nonparental ditypes and there are many tetratypes so we can conclude the *p* and mating loci are linked. RF = (NPD + 1/2T)/ total tetrads = (3) + (1/2)(27)/54 =16.5/54= .31 × 100 = 31 cM between the two genes.

4-21. a) The number of meioses represented here is the total of the number of asci formed. 334.
b) To map these genes, first designate the type of asci represented. This has to be done for each pair of loci: For *a* and *b*:

PD	NPD	T	T	NPD	T
a+c	*abc*	*++c*	*+bc*	*ab+*	*a+c*
a+c	*abc*	*a+c*	*abc*	*ab+*	*abc*
+b+	*+++*	*+b+*	*+++*	*++c*	*+++*
+b+	*+++*	*ab+*	*a++*	*++c*	*+b+*
137	141	26	25	2	3

Notice that the PD and NPD classes are almost equal indicating that these two genes are not linked. For genes *b* and *c*:

PD	NPD	PD	NPD	PD	T
a+c	*abc*	*++c*	*+bc*	*ab+*	*a+c*
a+c	*abc*	*a+c*	*abc*	*ab+*	*abc*
+b+	*+++*	*+b+*	*+++*	*++c*	*+++*
+b+	*+++*	*ab+*	*a++*	*++c*	*+b+*
137	141	26	25	2	3

Again the PD and NPD asci are about equal so genes *b* and *c* are not linked. For genes *a* and *c*:

PD	PD	T	T	NPD	PD
a+c	*abc*	*++c*	*+bc*	*ab+*	*a+c*
a+c	*abc*	*a+c*	*abc*	*ab+*	*abc*
+b+	*+++*	*+b+*	*+++*	*++c*	*+++*
+b+	*+++*	*ab+*	*a++*	*++c*	*+b+*
137	141	26	25	2	3

The number of PD asci is far greater than the NPD so the genes are linked.

$$\frac{2 + 1/2(26 + 25)}{334} = 8.2 \text{ cM between } a \text{ and } c.$$

There is second division segregation of a in two classes of asci (26 + 25) indicating the gene *a* is 1/2(51/334) or 7.63 cM from its centromere. The last ascus class shows second division segregation for *b* so *b* is 1/2(3/334) or 0.45 cM from a centromere. There is no second division segregation for c, so that gene is very close to the centromere.

$\bigcirc c \leftarrow 8.2 \rightarrow a$

$\bigcirc\ \ 0.45\ b$

4-22. The results fit the model that the *a* and *b* loci are on different chromosomes and are both very close to their respective centromeres, while the *c* locus is farther from its centromere. Tetratypes can arise in crosses between strains having markers on different chromosomes when a crossover occurs between one of the genes and its centromere. If neither locus experiences such a crossover, either a PD or an NPD ascus will be formed.

4-23. a) The two types of cells in the first group of 89 asci are *met$^+$lys$^+$* (could grow on all four types of media) and *met$^-$lys$^-$*. These asci are parental ditypes (PD). The four types of cells in the second group of 11 asci are *met$^+$lys$^+$; met$^+$lys$^-$* (grew only on min + lys); *met$^-$ lys$^+$; met$^-$ lys$^-$*. These are tetratype (T) asci.

b) Because the number of PD>>> NPD, the genes are linked. The distance between them is (NPD + ½ T) /total tetrads = (0 + ½(11))/100 × 100 = 5.5 m.u.

c) NPD should be seen and result from double crossovers involving four strands (chromatids). There would be two types of spores: met^-lys^+ and $met^+ lys^-$ these could grow on minimal + methionine and minimal + lysine respectively. (These two types of spores could also grow on min + met+lys.)

4-24. (Notice that this question asks for the number of different kinds of phenotypes- not the number of individuals with each of the different phenotypes.)

a) 2 b) 3 c) 3 d) 4 e) 4 (Note: Because the genes are linked, the four classes will not be equally represented.)

f) Three phenotypes are possible for each gene. The total number of combinations of phenotypes is $(3)^2$ or 9.

g) Normally there would be four phenotypic classes corresponding to genotypic classes *A-B-,*
A-bb, aaB- and *aabb* (From both recombination and no recombination between the linked genes) but since one of the genes is epistatic, two of the genotypic classes have the same phenotype so there are 3 phenotypic classes.

h) Four phenotypic classes corresponding to the genotypic classes *A-B-, aaB-, A-bb, aabb* are the four genotypic classes but the first three are all equivalent phenotypically so there are only 2 phenotypic classes.

i) In this case there is 100% linkage between the two genes. The number of phenotypic classes will be dependent on the arrangement of alleles in the individuals mating. If the parents are *AB/ab* x *AB/ab*, there will be two phenotypic classes. If the parents are *Ab/aB* × *Ab/aB*, there will be one phenotypic class produced in the offspring.

4-25. Red sectors would have to have an *ade2⁻/ade2⁻* genotype. These could arise by mitotic recombination.

4-26. 2, 5, 9 . After replication in the heterozygous diploid, each chromosome would be composed of a pair of sister chromatids (shown below). The centromeres on sister chromatids cause the chromatids to segregate from each other during mitosis. (Centromeres are given the chromatid number so you can follow them more easily). For example, CEN₁ and CEN₂ segregate from each other, so chromatids 1 or 2 and 3 or 4 will end up in a daughter cell.

Possible recombination locations between non-sister chromatids are indicated by X's below. (Assume that no crossovers occur between the *a* and the *CEN* since they are so close together.) The only way to get any of the possible phenotypes listed (which are homozygous recessive for one or more genes) would be in a cell that receives chromatid 1 (which will contain all recessive alleles) and a recombinant chromatid 3. Remember also that a cell homozygous for *leth* allele will die. Look at the genes that are connected to centromere in chromatid 3 after recombination at the various locations.

chromatid #

1   b a CEN_1 c *leth* d e

2   b a CEN_2 c *leth* d e
                X_1 X_2 X_3 X_4 X_5
3   b^+ a^+ CEN_3 c^+ *leth*$^+$ d^+ e^+

4   b^+ a^+ CEN_4 c^+ *leth*$^+$ d^+ e^+

X_1 produces $b\ a^+\ CEN_3\ c^+\ leth^+\ d^+\ e^+$ b genotype; lives
X_2 produces $b^+\ a^+\ CEN_3\ c\ leth\ d\ e$ dies
X_3 produces $b^+\ a^+\ CEN_3\ c^+\ leth\ d\ e$ dies
X_4 produces $b^+\ a^+\ CEN_3\ c^+\ leth^+\ d\ e$ $d\ e$ genotype; lives
X_5 produces $b^+\ a^+\ CEN_3\ c^+\ leth^+\ d^+\ e$ e genotype; lives

Chapter 5   How the Molecule of Heredity Carries, Replicates and Recombines Information

Synopsis

The statement "DNA's genetic functions flow directly from its molecular structure" is a good focus for reviewing DNA structure. By focusing on function, the beauty of the structure will become more evident. Make sure you really understand the structure and get a good mental image of the DNA molecule and its construction. Understanding where hydrogen and covalent bonds are found, the polarity of the strands of DNA, and why complementarity is important will provide a good basis for understanding many of the cellular processes (for example, transcription, translation, recombination) and the manipulations of recombinant DNA technology, from cloning to genetic screening.

You should become familiar with the evidence that replication is semiconservative and understand the general processes of initiation and elongation. Think about why special enzymes (such as topoisomerases and telomerases) are needed for replication of the genome. In the process of recombination, understand how heteroduplex regions are formed and what effect they have on recombination outcomes. You should work your way through the recombination process by drawing it out for yourself.

Be prepared to:

After reading the chapter and thinking about the concepts, you should be able to:
- assign the 5' and 3' designations to backbone strands of DNA and RNA
- indicate complementary bases to an RNA or DNA template
- explain the importance of structural features of DNA for function in copying (replication) in recombining information
- think about experimental design

Problem solving tips:

The nature of the chapter problems changes with this material. You will find the problems are more based on experimental design. The problems again require that you have a basic understanding of the concepts and use that knowledge. These problems can be viewed as more creative synthesis problems. A good way to approach these problems is to identify concepts relevant to the problem and review your knowledge of the topic and any relevant experimentation. The experimental type of question may cause you to go back and refresh your understanding of some techniques. One technique that comes up frequently is tagging a molecule with a label so the molecule can be followed. This is often done using radioactivity. Radioactive label can be incorporated into protein or DNA if a cell or organisms is grown in or fed a radioactive precursor that goes specifically into the type of molecule you want to follow. In designing experiments using radioactive labeling, be sure to consider how you can get a unique label into the molecule of interest.

Solutions to Problems

5-1. a) 6 b) 11 c) 9 d) 2 e) 4 f) 8 g) 10 h) 12 i) 3 j) 13 k) 5 l) 1 m) 7

5-2. The proof that DNA was the transforming principle was the treatment of the transforming extract with an enzyme (DNAse) that degrades DNA, and showing that the extract was no longer able to transform rough into smooth cells. (They also showed that treatments with RNAse and proteinase did not abolish the transforming activity of their extracts, indicating that the transforming principle was not RNA or protein.)

5-3. Gene *c* is closer to gene *a* because there are more instances when the two genes were cotransformed (and therefore were on the same fragment of DNA from the genome).

5-4. Although nitrogen and carbon are more abundant in proteins than the sulfur used by Hershey and Chase in labeling the bacteriophages in their experiment, nitrogen and carbon are not unique to proteins. Both elements are found in abundance in DNA as well, so there would be no way to differentiate protein and nucleic acid if labeled nitrogen or carbon compounds were used.

5-5. Tube #1, if all the sugar phosphate bonds are broken, would contain individual pairs of nucleotides. Complementary bases (bonded together still by hydrogen bonds) would each be attached to a sugar and phosphate. Tube #2 would contain base pairs (without the sugar and phosphate) and sugar phosphate chains without the bases. Tube #3 would contain single strands of DNA since the hydrogen bonds between bases were broken.

5-6. X-ray diffraction studies indicated that DNA was a helix containing repeating units spaced every 3.4Å. One complete turn of the helix occurs every 34Å. The diameter of the molecule, 20Å, indicated that DNA must be composed of more than one polynucleotide chain.

5-7. a) If 30% of the bases are A, and A pairs with an equal % T (30%), that leaves 40% to be C + G. There will be 20% C.
b) 30% T
c) 20% G

5-8. True statements : a) , b), e)

5-9. Double-stranded DNA contains complementary bases while single stranded DNA does not. The two forms of DNA can be distinguished by determining the % of each base. If A=T, C=G, the DNA is double-stranded. (An alternate experiment would be to treat the DNA with restriction enzymes. Double-stranded DNA is cut by restriction enzymes, single-stranded DNA is not.)

5-10. a) There are only two hydrogen bonds holding A-T base pairs together, whereas three hydrogen bonds hold the G-C base pairs together. It would therefore take less energy (heat) to break the bonds between A's and T's.
b) Regions of the denatured DNA must have contained sequences of bases complementary to a nearby sequence in inverted orientation. These regions of the molecule reformed a double-stranded region with a loop of single-stranded DNA between them.

5-11. 3' GGGAACCTTGATGTTTCGGCTCTAATT 5'

5-12. The experiment that was done to prove that RNA was the hereditary material in TMV was to mix the nucleic acid and protein from the viruses containing different proteins and see which type of protein was found in the progeny. RNA from virus type 1 was mixed with protein from virus type 2 to reconstitute a "hybrid" virus; RNA from phage type 2 was mixed with protein from virus type 1. When these reconstituted hybrid viruses were used to infect cells, the progeny had the protein that corresponded to the source of the RNA, not the protein, found in the hybrid virus.

5-13. 5' UAUACGAAUU 3'
(Primers for DNA synthesis are RNA molecules.)

5-14. c) Label would be in one chromatid each of both homologs. After one S phase, the label would be in one strand of each DNA molecule which means that each chromatid (a double-stranded molecule) would contain label. The label was then removed before the next S phase, so the second set of new strands would not be labeled. The copy made complementary to the unlabeled strand of each chromatid from the first S phase would also be unlabeled so that homolog would contain no label. The homologous chromatid would still contain a labeled strand from the first round of synthesis and the complementary copy would be unlabeled. Therefore, one chromatid of each homologous pair would contain label.

5-15. It is best to think about individual strands of DNA to calculate the amount of DNA of different densities. After two generations:

       HL   HL   and   LL   LL

After one additional generation:

HL   LL   HL   LL   LL   LL   LL   LL

1/4 double-stranded DNA molecules would be HL

After two additional generations:

HL LL LL LL HL LL LL LL LL LL LL LL LL LL LL LL

1/8 double-stranded DNA molecules would be HL.

5-16. There is one replication origin per bubble, but two replication forks. In this example there are three origins, and six forks.

5-17.

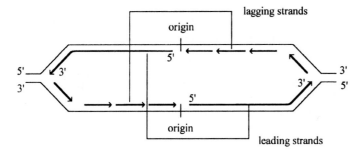

5-18.

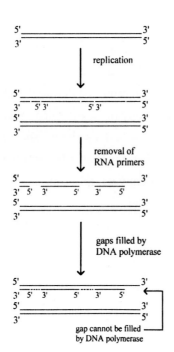

The problem at the end of a linear chromosome is that when the primer is removed from the 3' end, there will be no way to synthesize a DNA copy of the those exposed bases in the template. Information can therefore be lost from the end of the chromosome if there is not an alternate replication methodology at the ends.

5-19. a) Topoisomerase relieves the stress of the overwound DNA ahead of the replication fork. b) Helicase unwinds the DNA by breaking H bonds between base pairs to expose the single strand templates for replication.
c) Primase synthesizes a short RNA oligonucleotide which DNA polymerase will extend.
d) DNA ligase joins the sugar phosphate backbones of Okazaki fragments.

5-20. A break of both strands of the backbone of one of the intertwined molecules is necessary to unlock these two helices. The unbroken strand can then be passed through the break in the other DNA molecule.

5-21. The process of recombination involves the breakage, exchange, and rejoining of strands; and a final resolution by a cutting and rejoining of strands. If the same two strands of the DNA molecule that were cut during the initiating event are cut during the resolution, there will be no crossing over (genetic recombination of markers outside of the gene conversion region). There is an equal likelihood that the other strands will get cut and genetic recombination is the result. Regardless of which strands are cut during resolution (to result in crossing-over or no crossing-over), mismatches within the heteroduplex region generated early in the process can be corrected to the same allele, resulting in gene conversion. Therefore, gene conversion occurs equally frequently with and without recombination of genetic markers.

5-22. The numbers of B and b alleles are not in the 2:2 ratio predicted from reciprocal exchange during recombination. The 3:1 ratios indicate that gene conversion occurred. In formation of the spores in the two tetrads, the same type of recombination event occurred, generating a heteroduplex region including gene B but correction of the mismatch in the region of heteroduplex was different. In the first case the correction of mismatches in both duplexes was to B and the second case the correction in one duplex was to b and in the other correction was to B. In tetrad II there was also crossing-over between A and C.

5-23. A mutant *recA E. coli* strain would not be able to undergo recombination. Any event requiring recombination would not occur normally.

5-24. a) The crossover initiated either between e and f or f and g.

b) The resolving cut occurred between f and g or e and f (if you said the initiating cut was between e and f or f and g respectively in part a.). The heteroduplex region in this particular event must have included gene f to get the resulting octad.

c) The chromosome that ended up with the f^+ and f in the heteroduplex region was corrected by mismatch repair to f f. The e gene was not within the heteroduplex region and therefore there was no mismatch to repair and the alleles segregated 4^+:4^-.

Chapter 6   Anatomy and Function of a Gene: Dissection Through Mutation

Synopsis

This chapter is about MUTATIONS! They are the heart and soul of genetics-the basis of genetic variation, the raw material for evolution, an essential part of genetic analysis. Now that you know the structure of DNA, you can understand the molecular nature of different kinds of mutations; how errors arise that can result in mutation, and how errors can be corrected.

Two of the most important tools for genetic analysis are described: complementation and deletion mapping. Mutations are essential for these analyses.

In complementation analysis, gametes containing chromosomes with two different mutations are combined. If a gamete containing a mutation in gene A, but not gene B is combined with a gamete having a mutation in gene B but not A, gene products A and B will be made in the resulting zygote. (The mutations complement each other.) But if the mutations are in the same gene, there is no good copy of the gene available so the cell is still mutant. (No complementation occurs.) Complementation analysis is used to determine if mutations that result in the same phenotype are in the same gene. In this way, the number of genes required for a particular process can be determined.

Deletion analysis is used to determine the location and order of genes. It is a quicker way to determine map order than doing linkage analysis on each pair of mutations (or genes) along a chromosome.

The connection between genes and what the gene products do is apparent in the ordering of intermediate compounds and the enzymes that catalyze the conversion of one compound to another. The order of genes acting in an enzymatic pathway is determined based on the consequences of a defect (mutation) in a gene encoding a protein that is needed in the pathway.

Be prepared to:

After reading the chapter and thinking about the concepts, you should be able to:

- identify different types of mutations in a DNA sequence (frameshift, transition, transversion, deletion, inversion)
- know how to set up a complementation analysis
- determine the number of genes represented by a series of mutations by analyzing results of complementation tests
- use deletion mapping to position mutants on a map
- determine the order that genes act in an pathway composed of several enzymatic steps
- determine the order of intermediate compounds in an enzymatic pathway using data on the ability of mutants to grow on intermediate compounds

Problem Solving Tips

- Lack of complementation indicates that mutations are in the same gene. Complementation between two mutations means the mutations are in different genes.

- When two mutations complement each other, the wild-type phenotype is seen in all cells in which the two mutations combine. If there is not complementation, the cells will be mutant, unless recombination between the mutant chromosomes occurs to regenerate a wild-type gene. Recombination between the two mutations will occur at a low level, so if a small number of wild-type cells are seen, it is due to recombination (or reversion, if there has been no chance for recombination to occur)

- Complementation does not require any interaction between the DNA molecules; recombination does. Make sure you understand which process is being analyzed.

- Deletions do not revert and can be recognized by this characteristic. Deletions do not complement mutations that are located in the deleted region.

- Combining a gamete containing a deletion in one chromosome with a gamete containing recessive mutation(s) will expose the recessive mutant phenotype if those genes lie within the deleted region of DNA.

- An enzymatic pathway is a series of steps each catalyzed by a gene product (enzyme). Think through these problems logically. If there is a mutation early in the pathway, providing later intermediate compounds will allow growth, because the subsequent enzymes are available. The block point has effectively been bypassed in these cases.

Solutions to problems

6-1. 1) a,b,h   2) b,c   3) f,j   4) g,j   5) a,b,h   6) f,i,k   7) g, k   8)d,i   9) e,f,i

6-2. If you assume that a different single base change occurred to result in each of the three mutants, you can determine the wild-type sequence by finding the base that is present in two of the three sequences. A different base at one position will be the mutation.
The wild-type sequence is therefore:
ACCGTAGTCGACTGGTAAACTTTGCGCG

6-3. Of the 27 achondroplasia births observed, 23 are due to new mutations (no family history of dwarfism). Because achondroplasia is an autosomal dominant trait, it will be expressed immediately when it occurs. Since 120,000 births were registered, there were 240,000 parents in which the mutation could have occurred. The mutation rate can calculated directly:
$23/240,000 = 9.5 \times 10^{-5}$. This rate is higher than the 2 to 12×10^{-6} mutations per gene per generation calculated as the average mutation rate for humans.

c) Without crossovers, there would be mutant males that inherited the *y cv ct sn m* mutations and another class of males that contain a deletion (including the *ct sn* genes) and these flies die.

6-9. Yes. The rat liver supernatant contains enzymes that convert substance X to a mutagen, and his^+ revertants occur. Our livers contain similar enzymes that process substances, converting them into other forms that cause mutation and can lead to cancer.

6-10. Mutations in the tumor cells that arose after treatment with the mutagen are somatic cell mutations. When the tumor cells are injected they can cause a new tumor in a tissue but the mutant cells are not present in the germ cells (gametes). The cancer phenotype cannot be passed on by mating, because the mutation is not in the germ cells.

6-11. If there is a suppressor mutation present in another gene, the suppressor (su^-) and original met^- mutation should be separable (segregate independent of each other) during meiosis. If a $met^- \ su^-$ was mated with a wild-type strain $met^+ \ su^+$, the haploid progeny after sporulation of the diploid, would be $met^- \ su^-$; $met^- \ su^+$; $met^+ \ su^-$; $met^+ \ su^+$ in equal proportions. Phenotypically this would produce 3:1 met^+ : met^-. (This assumes that the *met* and *su* genes are not linked. If the genes were linked, you would still get the four genotypic classes but the proportions of the recombinant gametes would be dependent on the linkage between the genes.)

6-12. A complementation test could be performed by mating the two mice. If the mutations causing albininsm are in different genes, all the progeny should be wild-type. If the mutations are in the same gene, there would be no complementation and all progeny will be mutant (albino).

6-13. a) Mutations 5 and 6 are deletions (non-reverting mutations) so should not be considered here. Mutation 1 does not complement mutation 8. These make up one complementation group. Mutation 2 complements every other mutation, therefore it is alone as a complementation group. Mutation 3 does not complement 4,7 so these form a third complementation group. There are three complementation groups: 1,8; 2; 3,4,7
b) Recombination can occur in the diploid yeast strains between and within genes, producing prototrophic (lys^+) spores. If recombination occurs between nonsister chromatids, a tetrad will be composed of 3 lys^- and 1 lys^+ spore. (The numbers of tetrads showing this ratio will depend on the distance between the mutations.)

$$1^- \quad\quad 8^+ \qquad\qquad\qquad 1^- \quad\quad 8^+ \qquad lys^-$$

$$1^- \quad\quad 8^+ \qquad\qquad\qquad 1^- \quad\quad 8^- \qquad lys^-$$

$$\mathbf{X} \qquad \rightarrow$$

$$1^+ \quad\quad 8^- \qquad\qquad\qquad 1^+ \quad\quad 8^+ \qquad lys^+$$

$$1^+ \quad\quad 8^- \qquad\qquad\qquad 1^+ \quad\quad 8^- \qquad lys^-$$

c) From the recombination data in part b), mutants 5 and 6 seem unusual. Mutant 6 we concluded was a deletion and since it does not recombine with 1, 4, or 5 to produce any prototrophic spores (see table for section b), it must delete regions where these mutations map. Mutant 5 must also be a deletion because it does not recombine with 3, 4, or 6. From this deletion information and the conclusion in part a) that there are three genes shown on the map below:

```
2?  |  8   1   |   4   3   7 |   2?
----+----------+-------------+-----
    |  deletion 6 |
         |deletion 5 |
```

The location of gene 2 (to the right or left side)cannot be determined from this data.

6-14. a) There are two complementation groups and therefore two genes.
b) 1 and 4 are in the same complementation group; 2,3,5 are in another complementation group.

6-15. a) Mutants 3, 6, and 7 are deletion mutants (non-reverting). The second table gives results of an experiment in which recombination between mutant phages can occur. The positive growth (+) scored in the table is the result of recombination that occurred in the *E. coli* B host to generate wild-type phage that can now grow in *E. coli* K (λ). The - designation indicates that a few phage were able to form plaques. These are revertants. Point mutations can revert to wild-type, deletion mutations cannot. Three of the mutants did not form any plaques - mutants 3, 6, and 7, and these three must be deletions. (You can also tell that 6 is a deletion because it does not complement mutations that fall in two different complementation groups.)
b) The first table on the coinfection of *rII⁻* mutants into *E. coli* K (λ) gives results of complementation analysis and lets us place mutations in the two *rII* complementation groups. The two complementation groups (from the first table) are 1, 2 and 5; 4, 8 and 9. Use the deletions to order the mutations. Deletion 6 does not complement any of the mutants so must delete at least part of each of the two *rII* genes. It recombines with mutants 2, 5, and 9 to produce wild-type phage so must not delete DNA where mutations 2, 5, and 9 are located. Deletion mutant 3 also does not recombine with mutant 4 so it overlaps deletion 6. It does not

recombine with mutant 9 either so covers the region where mutation 9 is located. We can now place mutations 9, 4, 8 and 1 based on these deletions. Mutant 7 is deleted for the region covered by mutants 1, 2, and 5 and overlaps deletion mutation 6.

```
 (2,5)   1 |   8          4         9
         |      deletion 6       |
 |   deletion 7 |          |  deletion 3  |
```

c) The order of mutations 2 and 5 in *rIIA* cannot be determined form this data. To determine the order, you would need to use other deletions that occur in *rIIA* in recombination testing or cross mutants 2 and 5 and test for linkage to appropriate genetic markers in genes that lie to either side of the *rIIA/B* locus.

6-16. a) Recombination between the two *rosy* mutations in the heterozygous female generates the eight offspring with wild-type eyes. The eight progeny represent fertilization of one of the recombinant gamete classes. The other class (the reciprocal product of the recombination) would be $ry^{41} \; ry^{564}$ double mutation on the same chromosome. These we would not be able to differentiate form the single *ry* mutants. We assume they are found in equal numbers to the wild-type gamete produced by recombination (8). Recombination frequency is therefore 8 + 8 or 16/100,000 or .0016%. The distance between ry^{41} and ry^{564} is .0016 map units.

b) The wing and bristle phenotypes of those eight recombinant offspring are a consequence of the order of the ry^{41} and ry^{564} mutations and the *Ly* and *Sb* genes. Try both orientations to see which order produces wild-type eyes together with Lyra wings and stubble bristles.

i)
```
_____ ry41   Sb ___
_____
   Ly            ry564
```

ii)
```
_____ ry41   Sb
_____
   Ly     ry564
```

Orientation i) produces the recombinants obtained so the order must be $Ly \; ry^{41} \; ry^{564} \; Sb$.

6-17. a) A cross between *argE⁻* and *argH⁻* would produce a diploid *argE⁻/argE⁺ argH⁻/argH⁺*. When this is sporulated, the parental ditype asci (showing the genotype of four spore pairs) would

be: $argE^{-} \; argH^{+}$

   $argE^{-} \; argH^{+}$

   $argE^{+} \; argH^{-}$

   $argE^{+} argH^{-}$

All spores would be Arg-.

The non-parental ditype would be

$$argE^+ argH^+$$
$$argE^+ argH^+$$
$$argE^- argH^-$$
$$argE^- argH^-$$

Two spores would be Arg$^-$, two would be Arg$^+$ (Note: The assumption was made that the genes are unlinked since you were not told otherwise in the problem.)

For the first two spores of the parental ditypes (above), ornithine, citrulline, arginosuccinate or arginine could be supplied for growth. For the second two spores, no intermediates would allow growth, only arginine itself would allow growth. For the Arg$^-$ spores of the NPD ascus, only arginine would allow growth.

6-18. a) The F$_1$ snakes would be heterozygous. (O=wild-type, o = orange; L = wild-type, l = black). The genotype of the F$_1$ is therefore *OoLl*. The F$_2$ would have the genotypes and ratios 9*O-L-*, 3*O-ll*, 3*ooL-*, 1*ooll*. If orange and black are two intermediates in the pathway to brown (orange → black → brown) and the O gene product carries out the conversion from orange to black and the L gene product carries out the conversion from black to brown, the F$_2$ would have 9 brown, 3 black, 4 orange. In other words, epistasis would be seen. (Note: We don't know the order of orange and black in this pathway. If the order were black → orange → brown, a 9:3:4 ratio of brown, orange, and black would be seen.)

b) If there are two pathways, one producing orange and the other black, then there would be four different phenotypes in the F$_2$ generation. 9 brown (*O-L-*), 3 black (*ooL-*), 3 orange (*O-ll*), and one nonpigmented (*ooll*).

6-19. Designate the genes and alleles. The G gene product converts green to blue flowers, the mutant allele is g. Either of two gene products B or L can convert blue to purple flowers, b and l are the mutant alleles. The true-breeding green-flowered plant is *ggBBLL*; the true-breeding blue flowered plant is *GGbbll*. The F$_1$ plants are *GgBbLl* and the F$_2$ will be:

G-B-L-	3/4 × 3/4 × 3/4 = 27/64	purple
G-bbL-	3/4 × 1/4 × 3/4 = 9/64	purple
G-B-ll	3/4 × 3/4 × 1/4 = 9/64	purple
G-bbll	3/4 × 1/4 × 1/4 = 3/64	blue
ggB-L-	1/4 × 3/4 × 3/4 = 9/64	green
ggbbL-	1/4 × 1/4 × 3/4 = 3/64	green
ggB-ll	1/4 × 3/4 × 1/4 = 3/64	green
ggbbll	1/4 × 1/4 × 1/4 = 1/64	green

The ratio is 45 purple : 16 green : 3 blue.

6-20. Working from the final product backwards through the intermediates, look for the mutant which grows only when supplied with G. In this problem it is mutant 2. The mutation must be in the last enzyme preceding G. Then look for the mutant that grows only when supplied with G or one other intermediate. Mutant 7 can grow only when supplied with intermediate E or with G. This gives us two pieces of information. We now know the intermediate that precedes G and we know the gene that encodes the enzyme that converts an intermediate to E. In this way, continue working back through the pathway to get the answer.

mutant   6   1   5   3   4   7   2

pathway → F → D → A → C → B → E → G

6-21. a) This problem is worked in the same way as 6-20.

mutant   18   14   9   10   21

pathway   → D → B → A → C → thymidine

b) Double mutant 9 and 10 would accumulate intermediate B. 10 and 14 would accumulate intermediate D.

6-22. To solve this problem consider first only those mutants that are only defective in biosynthesis of one amino acid. The mutants defective in only the proline pathway are those that grow when given proline in the media but not when given glutamine. Mutants 2, 6, and 1 are this type. No intermediate will allow growth of mutant 2 so the defect must be in the final enzyme that produces proline. Working back from this point in the pathway, mutant 6 grows when supplied with intermediate A, so A is the final intermediate and 6 is blocked in the step that leads to A. Mutant 1 grows when supplied with intermediates E or A, indicating that E is prior to A in the pathway. For proline, so far we have

        1   6   2

      → E → A → pro

The same analysis is done for glutamine. Mutants 7 and 4 are defective in glutamine biosynthesis only and the order of enzymes and intermediates is:

        7   4

      → B → gln

Now look at the mutants that area defective in both glutamine and proline biosynthesis. Mutants 5 and 3 are of this type. Mutant 3 grows only if given intermediate C so must be blocked just prior to this step. Mutant 5 grows if given C or D so is blocked prior to the D intermediate.

        5   3

      → D → C

This represents the part of the pathway that is used both in proline and glutamine biosynthesis.

Putting this together with the branches that are specific for each amino acid, we have the following branching pathway.

$$5 \quad 3 \quad 1 \quad 6 \quad 2$$
$$\rightarrow D \rightarrow C \rightarrow E \rightarrow A \rightarrow pro$$
$$\downarrow 7 \quad 4$$
$$\rightarrow B \rightarrow gln$$

6-23.

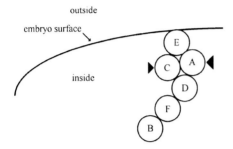

This problem is similar to an ordered enzymatic pathway except the gene products are a series of structural proteins that make up a structure and are dependent on assembly in a particular order. The loss of one protein due to mutation will prevent all the subsequent proteins from being added. The loss of the first protein at the surface would prevent all others from being at the cell surface. The mutant that fits this description is E. Mutants A and C have a similar pattern and show that only E and C or A respectively are at the surface so genes A and C encode the two proteins that form the dimer structure shown second from the embryo surface. The logic is continued to place the remaining three genes in their order.

6-24. Two loci are needed (one for the a and one for the b proteins).

6-25. In unequal crossing-over the homologous chromosomes align out of register using homology in this case between related but different genes.

The result of crossing-over is b d/b d (duplication of information) and b/d (deletion of information) where the genes with a slash indicate hybrid genes. Information critical to having a functioning b gene may have been lost from one chromatid.

6-26. a) Null mutations have no activity of the gene product. Hypomorphic mutations lead to a lower level of protein activity while hypermorphic mutations lead to a higher level of activity. Dominant negative mutations interfere with the functioning of normal protein (made by another allele) or interfere with the functioning of other proteins that interact with the gene product. Neomorphic mutations exhibit a new phenotype.

b) Null and hypomorphic mutations would be recessive to wild-type unless the phenotype is depended on the amount of a gene product. Hypermorphic and neomorphic mutations would probably be dominant. Dominant negative mutations are dominant by definition.

6-27. The problem states that the contributions of each allele are additive. The 100% level must be two wild-type alleles so each b^+ allele contributes 50% activity.

a)and b). Carol, with 50% activity is heterozygous for a normal allele and a null allele. (b^+b^0) Each normal allele contributes 50% activity. The 70% activity that Bill shows is due to a genotype of b^+b^{20} (50% + 20% activity from these two alleles)

c) There are four possible genotypes of a child of these parents: b^+b^+ (100% activity); b^+b^{20} (70% activity), b^0b^+ (50% activity), b^0b^{20} (20% activity). Only the last genotype, occurring with a 1/4 probability, leads to clinical thalassemia.

d) 1/4 probability of b^+b^+ to give 100% activity.

Chapter 7   Gene Expression: The Flow of Genetic Information from DNA via RNA to Protein

Synopsis

This chapter describes how information in DNA is converted into usable machinery (proteins) in the cell via the processes of transcription and translation. This flow of information is part of the central dogma of genetics. You need to become very comfortable with using the terms transcription and translation accurately. In transcription, DNA information is converted into RNA information. In translation, RNA information is converted into protein information. Work on developing some mental pictures for yourself – as if you can see the process occurring when you speak the words. The three letter DNA code and the correspondence between DNA sequence and protein sequence is described in this chapter. Tie together your knowledge of transcription/translation and the genetic code. This chapter contains many new vocabulary terms. The best way to know you have a good grasp on the terms is use the terms while pretending you are describing transcription, translation and the genetic code to another person.

Begin to introduce more inquiry into your learning process. For example, think about the components involved in transcription, RNA processing, and translation. How could they be affected by alterations (mutations) in any one of the components? Start thinking about how we know what we know and what evidence supports a particular view of how a process occurs.

Be prepared to:

After reading the chapter and thinking about the concepts, you should be able to:

- use a codon chart to do a virtual translation of DNA into protein sequence or reverse translate protein into DNA sequence
- identify open reading frames in a DNA sequence
- answer questions that require you to know the roles of the nucleic acids and proteins in transcription, translation, and processing
- be able to assign 5' and 3' end designations to DNA or RNA sequences
- identify the RNA-like strand in the double-stranded DNA sequence either by knowing the direction of transcription or by looking for open reading frames

Problem solving tips:

- Codons are found in mRNAs and anticodons are found in tRNAs

- Transcription occurs in a 5' to 3' direction (ribonucleotides are added to a 3' end)

- Information in the 5' portion of a coding region of the mRNA will be information in the NH_2 terminal portion of the protein.

- The convention is to write DNA sequences with the top strand running 5' to 3' left right.

Solutions to Problems

7-1. a) 5 b) 10 c) 8 d) 12 e) 6 f) 2 g) 9 h) 14 i) 3 j) 13 k) 1 l) 7 m) 15 n) 11 o) 4

7-2. a) 4 b) 6 c) 1 d) 2 e) 3 f) 5

7-3. a) GU GU GU GU GU   or   UG UG UG UG

b) GU UG GU UG GU UG GU UG GU

c) GUG UGU   GUG U etc (If you start with the second base instead of the first G, the codon sequence will be the same but shifted by one codon)

d) GUG UGU GUG UGU GUG UGU GUG UGU GU etc . This is the result of overlapping code starting on each new base consecutively. There are other possibilities reading for instance a codon starting on 1, 2, 4, 5, etc. The overlapping code will always give more coding information for the same number of bases compared to the non-overlapping code.

e) GUGU GUGU or UGUG UGUG (depends on where you start)

7-4. a) Comparing the mutant to the wild-type sequence, you can see where insertions (corresponding to a + mutation) and deletion (corresponding to a - mutation) occurred. The locations are noted on the figure below.

b) The amino acids in the wild-type and mutant protein are shown.

```
                   Lys  Ser  Pro  Ser  Leu  Asn  Ala
wild-type:      5' AAA AGT CCA TCA CTT AAT GCC 3'
                       (-)                   (+)
mutant:         5' AAA GTC CAT CAC TTA ATG GCC 3'
                   Lys  Val  His  His  Leu  Met  Ala
```

Five amino acids between the mutations are different from wild-type.

c) The substitutions of amino acids between the - and + mutations in the mutant must not alter the structure of the protein significantly enough to alter protein function.

7-5. Glutamic acid can be either GAA or GAG. In sickle cell anemia this amino acid is changed to valine by a single base change. Valine is encoded by GUX with X representing any of the four bases. The second base of the codon is altered in the Hb^S allele. In Hb^C the glutamic acid (GAA or GAG) is changed to a lysine (AAA or AAG). The change here is in the first base of the codon. Hb^C therefore precedes Hb^S in the map of β-globin gene (reading in the direction that the RNA polymerase travels along the gene).

7-6. The base sequence of the wild-type gene would be

GGX GCX CCX AGA AAA

G   G

CGX

The mutant is   GGX CAU CAA GGX AAA

C   G   G

Notice that there are several ambiguous bases (any one of four bases possible is indicated by X; other amino acids are encoded by two different codons) in the wild-type sequence. Proflavin induces frameshift mutations of single base insertions or deletions. By lining up invariant bases in the mutant with the wild-type, it is clear that a single base insertion occurred in the mutant. Knowing the amino acid sequence of the mutant and therefore the DNA sequence, all but one of the third base ambiguities in the wild type can be resolved. (The original sequence is shown in bold letters in the two sequences and is summarized below sequences.)

wild type   **GGX GCX CCX AGA AAA**

G G

CGX

mutant   GGX   CA UCA AGG XAAA

C   G   G

The third base in the first codon is ambiguous.

Deduced DNA sequence of wild-type:

5'   GGX GCA CCA AGG AAA   3'

7-7. Nierenberg and Leder determined that CUC is the leucine codon and UCU is a serine   codon using an in vitro translation system. The principle of the assay was that a synthetic triplet RNA codon, matching charged tRNA, and ribosome bound together would be too large to pass through a filter. They set up 20 reactions, each containing CUC, one radioactive amino acid attached to its tRNA, and the other 19 non-radioactive amino acids attached to their tRNAs. In the mixture containing the radioactive amino acid that corresponds to the codon CUC, radioactivity would be trapped on the filter. The same was done for the UCU triplet RNA.

7-8. a) If the Asn6 (AAC) is changed to a Tyr residue, the nucleotide change is to a UAC. In protein B this means that the Gln at position 3 becomes a Leu (CUA).

b) When Leu at position 8 is changed to Pro, the nucleotide change is from CUA to CCA. In protein B the Thr (ACU) at position 5 is still an Thr residue even though the codon is different (ACC).

c) When Gln at position 8 in protein B is changed to a Leu (CUA), the Lys codon (AAG) at position 11 in protein A is changed to a stop codon (UAG).

d) This is a thought question that involves some speculation. Once a protein coding sequence has evolved to produce a protein that forms a stable three-dimensional conformation, it is unlikely that a stable protein could be produced by the sequence shifted by one nucleotide. In addition, stop codons are usually encountered in other frames.

7-9. DNA sequences are generally written with the 5' to 3' strand on top; 3' to 5' strand on bottom. If the protein coding sequence for gene F is read from left to right, the RNA-like strand is left to right, and the template strand must be the bottom strand. The template for gene G is the opposite since the coding sequence is read in the opposite direction (right to left).The template is the top strand for gene G.

7-10. Genomic DNA contains intervening sequences or introns within a gene. These are not present in the mRNA that has been processed. Therefore there will be loops of non-hybridizing sequences that are from the genomic DNA.

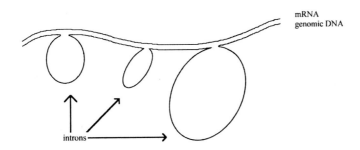

7-11. There are three possible reading frames to try on both the top and bottom strands. If a stop codon is found in a frame, there is not an open reading frame. If you scan the sequence looking for stop codons (TAA, TAG, or TGA), then determine which frames are open throughout the sequence, you find that there are two open reading frames in the top strand (reading from left to right) and one in the bottom strand (reading from right to left). The stop codons in the top and bottom strands are shown in bold.

5' CTTACAGTTTAT**TGA**TACGGAGAAGG 3'

3' **GAAT**GTCAAAT**AAC**TATGCCTCTTCC   5'

UGA   =   TGA
UAG   =   TAG
UAA   =   TAA

7-12. Eukaryotic genes contain intervening sequences (introns) that do not contain coding sequences. Introns do not have to contain open reading frames, therefore the open reading frame of a eukaryotic gene is often interrupted by sequence that will contain stop codons in the frame.

7-13. a) The nucleotide corresponding to the site at which the mutation in the DNA occurred is marked in bold and the consequences of the mutations are seen in the amino acid sequence below the RNA sequence.

normal	AUG ACA CAU CGA GGG GUG GUA AAC CCU AAG.......
	Met Thr His Arg Gly Val Val Asn Pro Lys
mutant 1	AUG ACA CAU C**C**A GGG GUG GUA AAC CCU AAG.......
transversion	Pro
mutant 2	AUG ACA CAU CGA GGG **U**GG UAA ACC CUA AG.......
deletion	Trp STOP
mutant 3	AUG AC**G** CAU CGA GGG GUG GUA AAC CCU AAG.......

transition	Thr (no change)

mutant 4   AUG ACA CAU CGA GGG GUU GGU AAA CCC UAA G.......

insertion   Val Gly Lys Pro STOP

mutant 5   AUG ACA CAU UGA GGG GUG GUA AAC CCU AAG.......

transition   STOP

mutant 6   AUG ACA UUU ACC ACC CCU CGA UGC CCU AAG

inversion   Phe Thr Thr Pro   Arg Cys

b) The single base transitions and transversions (mutants 1, 3, 5) could be reverted by EMS. The single base insertions and deletions (mutations 2, 4) could be reverted by proflavin.

7-14. First look at the sequence to determine where the Met Tyr Arg Gly Ala amino acids are encoded. On the bottom strand, there are Met Tyr codons on the far right (reading right to left) and Arg Gly Ala much farther down the same strand. What could separate these codons? There could be an intron. If there is an intron there should be splice donor and acceptor sites. These are present (marked below in bold).

a) The bottom strand is the RNA-like strand, so the top strand is the template. The RNA polymerase moves right to left along the template.

b) The sequence of the processed nucleotides (the joint between exons is marked by a vertical line) is:

        5'   AUG UAC AG|G GGG GCA UAG   3'

c) How can you get a Thr residue? This would occur if the G base on the bottom strand that followed the Tyr residue was mutated to a C. This is also a base change in the splice donor site, so splicing does not occur. The next codon after the Thr is a stop codon, so the polypeptide is only three amino acids long.

7-15. In transcription, complementary base pairing is required to add the appropriate ribonucleotide to a growing RNA chain. In translation, complementary base pairing between the codon in the mRNA and the anticodon in the tRNA is responsible for aligning the tRNA that carries the appropriate amino acid to be added to the polypeptide chain.

7-16. a) The minimum length of the coding region is 477 amino acids x 3 bases = 1431 base pairs (ignoring the stop codon). The gene could be longer if it contained introns.

b) Look at both strands of this sequence for the open reading frame. It occurs on the bottom strand, starting with the second base from the right. The direction of the protein is N-terminal to C-terminal going from right to left in the coding sequence.

5' GTAAGTTAACTTTCGACTAGTCCAGGGT 3'

3' CATTCAATTGAAAGGTGATCAGGTCCCA 5'

    Tyr Thr Leu Lys Gly Ser Thr Trp Pro

c) 5' ACCCUGGACUAGTCGAAAGUUAACUUAC 3'

7-17. a) The protein would terminate at the His codon because a nonsense mutation was present now after the His codon. The Trp codon (UGG) could have been changed to a UGA or UAG codon, both of which are stop codons. These stop codons result from a change in the second or third base of the Trp codon to an A.

b) To determine where else a single base change could produce a stop codon, write out the codons possible for the amino acid sequence.

    Ala Pro His Trp Arg Lys Gly Val Thr

    GCX CCX CAU UGG CGX AAA GGX GUX ACX

           C   AGG

If the Arg codon is CGU, a single base insertion in the DNA before or within this codon would lead to a UAA codon in the RNA. (The U would be from the former Arg codon and AA from the Lys codon.) If there was a mutation A to T at the position corresponding to the first base of the Lys codon, a UAA would be produced. If the Gly codon is GGA, a mutation of G to T corresponding to the first base of the Gly codon would generate a nonsense codon (UGA). If the Val codon is GUG, a deletion of the first base would uncover a stop codon.

7-18. The extra amino acids could come from an intron that is not spliced out due to a mutation in a splice site. The genomic DNA sequence in normal cells should contain this sequence. Another alternative is that the extra amino acids could come from an insertion of DNA from some other part of the genome.

7-19. a) The sequence of the anticodon is complementary to the codon. The anti-codon in this suppressor tRNA is 3' AUC 5'

b) The wild-type tRNA^{Gln} has to recognize a CAA or CAG codon. Since only a single base change occurred to form the suppressor tRNA that recognizes a UAG codon, the starting tRNA must have been recognizing the CAG codon. The anticodon sequence in the wild-type tRNA is therefore 3' GUC 5'. The template strand used to produce the tRNA is complementary to the RNA sequence itself. The sequence of the template (RNA-like) strand is 5' CAG 3'.

c) There must be a minimum of two genes. One would be the gene that was changed into a nonsense suppressor, and the other gene would have to be able to recognize all the normal Gln codons (5' CCA 3' or 5' CAG 3'). One tRNA could recognize either of these Gln codons based on the wobble in pairing at the 3' end of a codon.

7-20. a) The wild-type gene would be 5'CCG 3'. Because there is a progression of mutations from Pro (CCX) to Ser (TCX) to Trp (TGG) codons, the original proline codon must have been CCG.

b) Because no protein is made, the mutation is strain C is probably a nonsense mutation and it would be a 5' TAG 3' stop codon. The wild-type GGC was mutated to 5' TGG 3' in strain B, then to 5' TAG 3' in strain C.

c) Mutation in strain C-1 is a nonsense suppressor- a mutation in a tRNA that inserts an amino acid at UAG nonsense codons.

7-21. A missense suppressing mutation has the potential to change the identity of a particular amino acid inserted in many normal proteins (without the missense mutation) whereas a nonsense suppressing mutation can only make a normal protein longer.

b) You could have a mutation i) in a tRNA gene that would change the anticodon to recognize a different codon, ii) in a tRNA gene in a region other than the anticodon itself so that the wrong aminoacyl synthetase would sometimes recognize the tRNA and put the wrong amino acid on it, iii) in an aminoacyl synthetase gene making an enzyme that would sometimes put the wrong amino acid on the tRNA, iv) in a ribosomal protein, ribosomal RNA or translation factor that would make the ribosome more error-prone, putting the wrong amino acid in the polypeptide, v) mutation in the RNA polymerase gene that would sometimes transcribe the sequence incorrectly.

7-22. In the second bacterial species where the isolation of nonsense suppressors was not possible, there must be only a single tRNA-Tyr and only a single tRNA-Gln gene. Thus, neither gene could mutate to a nonsense suppressor, because there would not be any tRNA that could put Tyr or Gln where they belong. In this scenario, the single tRNA Tyr would have to have an anticodon of 3' AUG 5' to recognize the two Tyr codons of 5' UAU 3' and 5' UAC 3' by wobble. The single tRNA Gln would have to have an anticodon of 3' GUU 5' to recognize both 5' CAG 3' and 5' CAA 3' Gln codons by wobble.

7-23. Mutations near the C-terminus are expected to be less severe than N-terminal mutations because there is less of the protein that will be affected. Even nonsense mutations near the C-terminus can be tolerated in some cases.

a) very severe

b) probably mild effects

c) very severe

d) probably mild effects

e) no effect

f) mild to no effect

g) severe

h) could be severe if it affects the protein structure enough to hinder the action of the protein, or could be mild.

7-24. a) Promoter down mutations: no RNA produced.

b) Splice donor and acceptor mutations would result in different proteins (or no protein produced if a nonsense codon is uncovered).

c) No translation would occur unless a secondary binding site is used.

d) Only the splicing mutations still have the potential to produce protein and therefore could interfere with normal proteins in a dominant negative way.

Chapter 8   DNA at High Resolution: The Use of DNA Cloning, PCR, and Hybridization as Tools of Genetic Analysis

Synopsis

This chapter introduces you to many of the recombinant DNA techniques that have provided a powerful new approach for studying the mechanisms of inheritance and functions of specific genes. Restriction enzymes, cloning DNA, making libraries, identifying clones of interest, DNA sequencing, PCR amplification are now just a part of the toolkit that all biologists (not just geneticists) use. These techniques will be referred to over and over throughout this textbook (and probably in your other biology courses as well) so it is worthwhile to get a solid understanding of these techniques from this chapter.

As you read about the various techniques and apply them to solve problems, try to keep in mind which techniques are done in solutions in test tubes (restriction enzyme digests, ligating fragments together, PCR) and which techniques involve analyzing or manipulating DNA in cells (transformations, screening libraries, preparing large amounts of total genomic DNA or cellular RNA). This should help your understanding of the techniques and their uses. Hybridization of nucleic acids is central to many techniques but is often challenging to understand. The basis of hybridization is *complementarity* of bases in forming double stranded nucleic acids. A probe DNA or RNA molecule is used to locate a specific sequence (in a gel or as a clone inside a cell) based on hybridization. A probe contains a recognizable radioactive or fluorescent tag that makes it possible to identify the place where the probe found a complementary sequence.

Be prepared to:

After reading the chapter and thinking about the concepts, you should be able to:

- make a map of restriction enzyme sites given information on sizes of fragments from single and double digests
- describe the basic components and uses of different types of cloning vectors
- describe the essential steps in cloning
- read and interpret DNA sequencing gels
- design PCR primers

Problem solving tips:

- Be prepared to go through some trial and error in solving restriction map puzzles. Use pencil and eraser. Be patient. To start, try to find the clearest results that allow you to place some sites unambiguously. Once you have placed sites for two different enzymes on a map, sites for any other enzymes will have to fit in accordance with that initial map you make. Make sure the final sites you put on a map are consistent with results from all digests.

- Sticky ends that have complementary overhanging single-stranded bases can be ligated together. It may be helpful to draw out the 5' and 3' ends generated (including the individual bases of the recognition site) when a double stranded DNA is cut by a restriction enzyme.
- cDNAs are generated starting with RNA molecules.
- cDNA libraries of clones contain only the regions of genes that are present in processed (spliced) transcripts synthesized in the cell from which they were isolated.
- Genomic libraries of clones contain all of the DNA (genes and non-coding regions) from the cells.
- Use your knowledge of requirements for transcription and translation when thinking about whether genes cloned into expression vectors will be expressed in the host cell.
- Insertion of a fragment into the middle of the *lacZ* gene inactivates the gene and a functional β-galactosidase enzyme will not be produced. Cells carrying a insert within the *lacZ* gene are unable to cleave a lactose-like substrate and are phenotypically Lac⁻. They are recognized as white colonies if the media contains the substrate X-gal.

Solutions to Problems

8-1. a) 10 b) 1 c) 9 d) 7 e) 6 f) 2 g) 8 h) 3 i) 5 j) 4

8-2. a) With a restriction recognition site that is 4 bases long, Sau3A recognition sites are expected every 4^4 or 256 bases. Since the human genome contains about 3×10^9 bases, one would expect $3 \times 10^9 / 256 = 1.17 \times 10^7$ (or about 12,000,000) fragments.

b) BamHI recognition sites (6 bases long) would be expected every 4^6 or 4096 (about 4100) bases. $3 \times 10^9 / 4100 = 7.3 \times 10^5$ (or about 700,000) fragments would be expected.

c) SfiI has a recognition site that includes 8 specific bases. The N indicates that any of the four bases is possible at that site and therefore does not enter into the calculations. Recognition sites would be expected every 4^8 or 65,536 (about 65,500) bases. $3 \times 10^9 / 65,500 = 4.6 \times 10^4$ (about 46,000) fragments would be expected.

8-3. Selectable markers in vectors are genes that provide a means of determining which cells in the transformation mix received the vector. They are often drug resistance genes so a drug can be added to the media and only those cells that have received and maintained the vector will grow.

8-4. Plasmids contain an origin of replication, a selectable gene and convenient restriction sites into which fragments can be cloned. Fragments of sizes up to 15 kb can be cloned into plasmids. Cosmids are hybrids between plasmids and bacteriophage lambda and contain a plasmid origin of replication, selectable gene, convenient restriction sites, and the cohesive ends of lambda that allow them to be

packaged into bacteriophage lambda particles. Up to 45 kb of DNA can be inserted into cosmids.
YACs contain yeast origins of replication, centromeres, telomeres, and yeast selectable genes.
Fragments of up to 1 Mb can be cloned into YAC vectors.

8-5. To answer this question, work through the results of ligation of the DNA fragments and the
vector. The vector is cut with BamHI which will leave the following ends:

......G   GATCC.......
......CCTAG   G.................

When an MboI fragment is cloned into the site, the site may not be cleavable by BamHI anymore.
That will depend on the base sequence at the ends of the fragment. The X in the sequence below
indicates the ambiguity of that base. In all cases the following sequence will be found: (The
sequences from the inserted fragment are in bold).

.....G**GATCX**.....................**X**GATCC...........
......**CCTAGX**.....................**XCTAG**G.........

a) 100% of the junctions will be cleavable by MboI

b) For a junction to be cleavable by BamHI there must be a C in the sixth position. This would occur
1/4 times or 25% of the time.

c) None of the junctions will be cleavable by XorII.

d) The first five bases fit the recognition site for EcoRII. The final position must be a pyrimidine (C
or T). There is a 1/2 chance that the junction contains the EcoRII site.

e) For the cleavage site to be a BamHI site in the human genome, it must have had a G at the 5'
position in the human DNA. (This G is contributed by the vector in the clones created.) Of the
BamHI sites being considered in this question, the human fragments have the C at the 3' end of the
recognition sequence because they are cut by BamHI when cloned into the vector, so that base is not
considered here. 3/4 of the fragments would have had another base other than G and would therefore
not be BamHI sites.

8-6. a) The genomic library is the most inclusive of all these libraries with all chromosomal
sequences represented. It would therefore consist of the greatest number of different clones.

b) All of these libraries would overlap each other to some extent. The genomic library contains all
the DNA sequences, the unique library will contain a subset of the genomic DNA, so they share
sequences. The cDNA libraries will each contain some sequences that do not overlap with each
other, but there are also common genes expressed in both brain and liver that would result in some
overlap of clones. The cDNAs contain some of the genomic sequences although intron DNA
sequences that separate exons will not be present as they are in a genomic library.

c) The starting material for a genomic library is all the DNA inside the cell. The cDNA libraries start
from RNA present in the cells (and represent therefore the expressed genes in these cells). For the
unique library, repetitive sequences would have to be removed.

8-7. a) Five genome equivalents are needed to reach a 95% confidence level that you will find a particular unique DNA sequence.

b) To determine the number of clones you have to screen to have looked through five genomic equivalents, first determine the number of clones in one genome equivalent. Divide the number of base pairs in the genome by the average insert size, then multiply by five to get the number of clones in five genome equivalents.

8-8. a) An intact copy of the whole gene would be on a fragment greater than 140 kb and would therefore have to be cloned into a YAC vector.

b) The entire coding sequence is 38.7 kb and could be cloned into a cosmid vector starting with a cDNA copy of the gene.

c) Exons are usually small enough to clone into plasmids. If this exon was less than 15 kb it could be cloned into a plasmid.

8-9. Order of steps: d; g; e; h; c; f; b; a. There is some flexibility in the order. Step d) must precede g); but these two steps could occur before or after e) and step b) could be done anytime before a).

8-10. a) A vector that is able to be transformed, selected, and maintained in animal cells and *E. coli* must contain a gene for selection in *E. coli* and a gene for selection in animal cells and replication origins for each type of cell.

b) To get expression of a human gene in *E. coli*, the cloning site should be next to sequences containing bacterial transcription and translation signals so the protein will be expressed in *E. coli*.

8-11. Longer DNA molecules take up more volume and therefore bump into the gel matrix, slowing down the molecule's movement. Shorter molecules can easily slip through many pore sizes in the gel matrix.

8-12. Because one fragment was generated with each of the two enzymes, BamHI and EcoRI each cut the plasmid once. The double digest gives information about the relative positions of these two sites. Viewing the sites in one direction the EcoRI site is 3 kb away from the BamHI site. (Viewed in the other direction, the EcoRI site is 6kb from the BamHI site.)

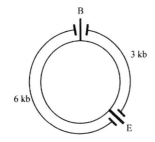

8-13. The fragment was cloned into the vector at the EcoRI site, therefore after digestion of the recombinant plasmid with EcoRI, the small EcoRI fragment of 49 bp and the vector fragment (900 + 500 + 300 + 700) = 2400 bp long would be generated. To predict the MboI digestion results, look through the insert fragment for the MboI sites. The inserted fragment contains an MboI recognition site, 5 bp into the fragment. There are two possible orientations in which the insert fragment could ligate. For one orientation, digestion with MboI produces 700 + 5 (705) bp, 44 + 900 (944) bp, 500 bp and 300 bp fragments. In the other orientation, the MboI digest produces a 900 + 5 (905), 700 + 44 (744), 500 and 300 bp fragments.

8-14. Draw the recombinant plasmid to help you determine the fragment sizes before sketching the gel.

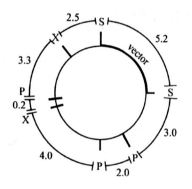

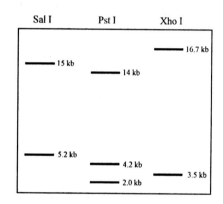

8-15. a) To think about which digest represents a linear versus a circular plasmid digest, consider the following simple case. What difference is there in products between a circular and linear molecule containing one restriction site? A restriction digest of the circular molecule results in one, while the digest of a linear molecule generates two fragments. Digestion of a circular molecule will always result in one fewer restriction fragments than the digest of a linear molecule. Sample A is therefore the circular form of the bacteriophage DNA.
b) The length of the linear molecule is determined by adding the lengths of the fragments from one digest. 5.0 + 3.0 + 2.0 kb = 10.0 kb. (This size is not realistic - λ DNA is, in fact, about 50 kb in length.)
c) The circular form has the same length: 10.0 kb.

d) Comparison of the circular and linear maps gives you information on which fragments contain the ends of the linear molecule. The 5.0 kb EcoRI fragment is present in the circular but not the linear digest so the 4.0 and 1.0 kb fragments must be at either ends of the linear molecule. (Start a working map of the molecule for yourself at this point.) The 2.7 kb BamHI fragment is present in the circular but not the linear digest so the 2.2kb and 0.5 kb (that total up to 2.7 kb) must be at the ends of the linear molecule. If the 0.5 kb BamHI fragment was at the end where the EcoRI 1.0 kb fragment is, the 1.0 kb EcoRI fragment would have been cut by BamHI in the double digest. However, the 1.0 kb fragment is still in the double digest, so the BamHI site 0.5 kb from an end must be at the end containing the 4.0 kb EcoRI fragment. The remaining EcoRI site is placed based on the double digests. The 2.0 kb EcoRI fragment is not cut by BamHI but the 3.0 kb fragment is, so place the site within the 3.0 kb. Now double check that all the BamHI-EcoRI fragment sizes are as seen in the different double digests.

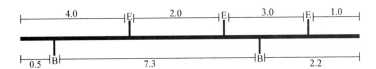

8-16. There are many different ways to start to solve restriction maps. The EcoRI and SalI digests indicate that there is one site for each of these enzymes. HindIII, in contrast, cuts the molecule at three sites. Draw a circle and place these three HindIII sites. When the plasmid was digested with both SalI and HindIII, the 2.0 and 1.0 kb fragments seen in the single HindIII digest are still present but the 4.0 kb HindIII fragment was cut into 2.5 and 1.5 kb fragments. The SalI site is therefore found in the 4.0 kb fragment. Place the SalI site with respect to HindIII sites. Similarly the EcoRI and HindIII double digest results in splitting the 1.0 kb HindIII fragment, but the orientation of the sites within the 1.0 kb HindIII is ambiguous because we have to consider relationships to the SalI site. Go to the EcoRI and HindIII digest. Try placing the EcoRI site in two different orientations to generate the 0.6 and 0.4 kb products. Then see how this fits with the EcoRI-SalI digestion results. The orientation that works is to place the 0.4 kb Hind-EcoRI fragment adjacent to the 2.5 kb SalI-HindIII fragment.

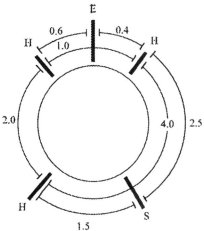

8-17. Enzymes EcoRI, HindIII and PstI each cut the plasmid once. BamHI cuts two times. Start your map by placing a BamHI site to create the 4.5 and 2.5 kb fragments. What BamHI fragment is cut in the BamHI + EcoRI double digest? The 4.5 kb fragment is missing and therefore is cut by EcoRI. What BamHI fragment is cut in the BamHI + HindIII double digest? The 2.5 kb fragment is cut. Try placing the EcoRI and HindIII sites within the BamHI fragments so that the sites are consistent with the double digest results. The 2.0 kb HindIII-BamHI must be adjacent to the 3.0 kb BamHI-EcoRI fragments to generate a 5.0 kb HindIII-EcoRI fragment in the double digest. PstI must be right next to EcoRI because there is only one band seen in the double digest with PstI + EcoRI and the band is the same size as the Pst or EcoRI digests alone. Consistent with this designation is the fact that the PstI + BamHI digest looks just like the EcoRI + BamHI digest.

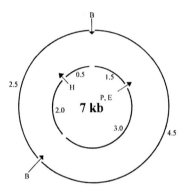

8-18. A partial digest results in all possible combinations of adjacent fragments as well as each fragment alone.

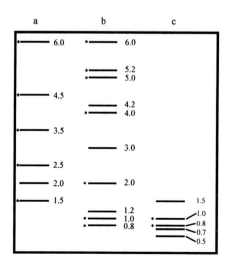

a) A partial digest with EcoRI:1.5, 2.0, 2.5, 3.5, 6, 4.5. The 1.5, 3.5, and 6.0 kb fragments contain the left terminus. The 2.5, 4.5 and 6.0 contain the right terminus. Terminal fragments are indicated with astericks.

b) A partial digest with XhoI: 0.8, 1.2, 3, 1, 2.0, 5, 6, 3.2, 5.2, 4. The 0.8, 2.1, 5.0, 6.0 fragments contain the left terminus. The 1, 4, 5.2, and 6 kb fragments contain the right terminus.

c) The complete double digest from left to right with XhoI and EcoRI results in 0.8, 0.7, 0.5, 1.5, 1.5, 1.0. The 0.8 kb fragment contains the left terminus and the 1.0 kb fragment contains the right terminus.

8-19. a) The full length is represented by the largest of the partial digest fragments- in fact this is not a partial at all, but one that has not been cut even once. It is 1.83 kb in length.

b) Because the figure shows the results of autoradiography, each of the sites starting from the labeled end are represented by a fragment in the gel. Subtract each subsequent restriction fragment size from the preceding one to get the sizes of fragments. For HhaI (in order from left to right): 0.42, .09, 0.93, 0.26, 0.13. For SalI: 0.26, 0.64, 0.35, 0.10, 0.24, 0.24.

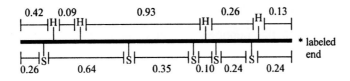

c) The complete double digest will yield the following fragments: 0.26, 0.16, 0.09, 0.39, 0.35, 0.10, 0.09, 0.15, 0.11, 0.13. All of these fragments will be seen on an ethidium bromide stained gel. Only the fragment that includes the labeled EcoRI end will be seen with autoradiography. The 0.13 kb fragment is rightmost on the map and contains the labeled end.

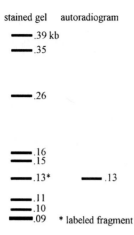

stained gel   autoradiogram

━ .39 kb
━ .35

━ .26

━ .16
━ .15
━ .13*   ━ .13
━ .11
━ .10
━ .09   * labeled fragment

8-20. 1) 3.1, 6.9 kb 2) 4.3, 4.0, 1.7 kb 3) 1.5, 0.6, 1.0, 6.9 kb 4) 4.3, 2.1, 1.9, 1.7 kb 5) 3.1, 1.2, 4.0, 1.7 kb

b)The 6.9 kb fragment in the EcoRI-HindIII digest; the 2.1 and 1.9 kb fragments in the BamHI-PstI, and the 4.0 kb fragment in the EcoRI-BamHI digest will hybridize with the 4.0 kb probe.

8-21. a) If the gene has been isolated from other organisms, you could use a probe of the gene from another organism. If the protein sequence of the ozonase enzyme is known, the amino acid sequence could be "reverse translated" into potential DNA sequences that could encode that protein. A mixture of oligonucleotides that could encode that peptide sequence could be used as a probe to hybridize to a library of clones.

b) Because the enzyme PstI generates only one fragment, the fragment length represents the total plasmid size. From the diagram, we see that the vector size is 5.5 kb. The insert is therefore 12.5 kb - 5.5 kb or 7 kb.

c) The map below is a representation of a linearized form of the entire plasmid. The insert portion is shown on the left side. The BamHI digest shows the insert (7.0 kb) and vector (5.5 kb) fragments. EcoRI cuts the plasmid three times. Two of the EcoRI sites are in the vector (refer to the diagram of the vector) so the one additional site must be in the insert. Look at the BamHI-EcoRI digest to place the EcoRI site within the BamHI fragment. The 7.0 kb BamHI fragment is cut into 5.5 and 1.5 kb fragments by EcoRI. Because there is a 2.0 kb fragment in the EcoRI only digestion, the 1.5 kb BamHI- EcoRI fragment must be adjacent to the 0.5 kb BamHI-EcoRI fragment. HindIII cuts the 7.0 kb insert into 5.0 and 2.0 kb fragments. Use the EcoRI-HindIII digest to determine the orientation. The 6.5 kb EcoRI fragment is cut into 3.5 and 3.0 kb fragments. The BamHI-HindIII

2.0 kb fragment must be closer to the left end. For PstI, the 7.0 kb BamHI fragment is cut into 4.5 and 2.5 kb fragments with the additional PstI cut. Place the PstI site in relation to the EcoRI sites using the EcoRI-PstI digest. Confirm the placement of sites using the triple digest results.

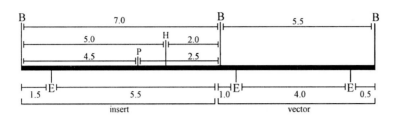

8-22. a) The strand that is synthesized is complementary to the template strand. Reading the sequencing gel from smallest to largest fragment, the sequence is:

synthesized strand: 5' TAGCTAGGCTAGCCCTTTATCG 3'

template strand: 3' ATCGATCCGATCGGGAAATAGC 5'

b) If the DNA sequence represents an exon, it should not contain stop codons. Scan the template strand (which you were told resembles the mRNA strand) in all three frames for stop codons. (Remember to scan in the 5' to 3' direction.) There are stop codons in each frame so it is unlikely that this is an exon sequence.

8-23. First read the DNA sequence from the gel. From the shortest to longest fragment seen, the sequence of the synthesized strand is:

TAGATAAGGAATGTAAGATATAACTGAGATTTAAC

The corresponding mRNA sequence is:

UAGAUAAGGAAUGUAAGAUAUAACUGAGAUUUAAC

The single base change in the gene (and therefore in the mRNA) is underlined.

8-24. a) Synthesis occurs in the 5' to 3' direction, so the smallest fragment would contain the 5' nucleotide added to the primer and the next sized product would be the TC.

8-24. a) Synthesis occurs in the 5' to 3' direction, so the smallest fragment would contain the 5' nucleotide added to the primer and the next sized product would be the TC.

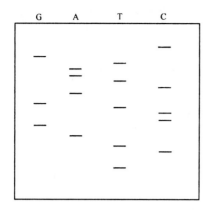

b) First write out the sequence of both strands and scan each strand for stop codons. The newly synthesized strand has stop codons in all three frames (underlined) and therefore would not be the coding (exon) sequence. On the template strand,

5' GCATTAGTTCAGGCTAC 3',

one of the reading frames beginning with the first three bases does not contain a stop codon and therefore is the frame read in this RNA-like strand.

Synthesized strand: 5' TCTAGCCTGAACTAATGC 3'

Template for DNA sequencing: 3' AGATCGGACTTGATTACG 5'

c) The peptide sequence beginning with the amino terminal end (which corresponds to the 5' end of the DNA sequence) is Ala-Leu-Val-Gln-Ala-Thr-Arg

8-25. Primers have to be 5' to 3' and have the 3' end toward the center so DNA polymerase can extend into the sequence being amplified. Set b) satisfies these criteria.

Chapter 9   The Direct Detection of Genotype

Synopsis

This chapter deals with a familiar genetic theme: variation. The variation described in this chapter is not the easily visible phenotypes discussed in early chapters, but molecular variation – changes in DNA sequence that can be detected by several different molecular techniques. When DNA is examined, the amount of variation (changes in sequence of bases) between individuals and among individuals in a population is much greater than you may have realized. Differences in DNA sequence seen in a population are called polymorphisms. These can be single bases changes or changes in the number of copies of a small repeated sequence (mini or micro satellites). If the single base variation changes a restriction enzyme recognition sequence, the polymorphism can be identified by restriction enzyme digestion of genomic DNA and hybridization to visualize only that region of DNA. Increases or decreases in repeated DNA can be recognized by changes in the size of restriction fragments or of PCR amplified regions. Changes in base sequence that do not affect a restriction enzyme recognition site or the size of a restriction enzyme fragment can be recognized using allele specific oligonucleotides and hybridization.

Much of the DNA variation seen between individuals is silent. That is, it has no effect on phenotype. A molecular genotype detailing the form (alleles) of DNA markers present in an individual can be used to distinguish between individuals and to map disease genes. The concept of linkage that you learned in chapter 4 is very critical again here. Genes and molecular markers that segregate (are transmitted) together greater than 50% of the time are linked. They are physically close on a chromosome and will only be separated if recombination occurs between them. Polymorphic DNA markers can be used to follow the transmission of a disease locus if the marker is linked to the disease gene.

Be prepared to:
After reading the chapter and thinking about the concepts, you should be able to:
- describe different types of DNA polymorphisms and the techniques used to detect the different polymorphisms
- distinguish different forms of a DNA marker and assign heterozygote and homozygote genotypes
- determine if a particular marker is informative for mapping a disease gene (Is it linked and is it polymorphic in the family you are studying?)
- follow inheritance of a disease allele of a gene using a linked molecular marker
- determine the likelihood that a particular individual carries a disease allele based on the molecular marker data and the % linkage between the marker and the disease gene

Problem-solving tips

- Remember that DNA markers do not have to be (and in fact usually are not) in a disease gene that is being traced. Instead the markers are DNA polymorphisms that are near a disease gene and are therefore linked.
- RFLP alleles are also referred to as forms of a polymorphism
- In an RFLP problem, first establish what different forms (alleles) are present in the population being considered
- A form (allele) of a DNA marker and the allele of a linked gene are transmitted together in a family unless there is recombination.
- A different allele of a marker can be associated with the same disease locus in two different families.

Solutions to Problems

9-1. a) 5 b) 3 c) 8 d) 6 e) 2 f) 7 g) 1 h) 4

9-2. The base difference could be a polymorphism that is not within the coding region of the gene. There are, on average, single base differences between two individuals every 500 bases.

9-3. a) Polymorphisms in microsatellites are different numbers of the simple sequence repeats.
b) Changes in the number of repeats are caused by slippage of DNA polymerase during replication.
c) Polymorphisms in minisatellites occur by unequal crossing-over if sequences on homologs align out of register during mitosis or meiosis. The repeat sequence in a minisatellite is too long to cause slippage of DNA polymerase during replication.

9-4. The gene detected by this clone must contain a polymorphism that alters the recognition site of the restriction enzyme used to digest this DNA.

9-5. a) A good way to determine the number of alleles is to find an individual who has only one band. Such an individual is homozygous if the gene is autosomal and that single band represents one of the RFLP alleles. Tim shows only one band and therefore one allele- the 3 kb version. Dad and mom each have a copy of 3 kb allele. The two additional bands present in dad's DNA represent a second RFLP allele and the additional band in mom's DNA represents a third allele. There are three different alleles represented in this family: 3 kb; 6 kb; and 4+2 kb form.
b) The dad is heterozygous for the 3 and 4+2 kb alleles. The mom is heterozygous for the 6 and 3 kb alleles. Sy is heterozygous for the 6 and 4+2 alleles. Tim is homozygous for the 3 kb alleles.

9-6. a) There are four alleles of the RFLP seen. To draw a map of the region, start with the largest fragment detected: 8 kb. The 4 kb allele found in sample 7 must represent the 8 kb fragment cut in two. The 7 kb fragment can arise if there is an additional EcoRI site 1 kb from the end of the fragment. Because the 1 kb fragment is not seen in the Southern analysis (on the left side of the fragment as drawn below), the probe used to detect these fragments must not include DNA sequences in this region. The 4+3 kb allele arises if there is an additional EcoRI site within the 7 kb region.

```
E 1.0 E   3.0      E      4.0       E
|_____|_____|_____|
```

b) There are two variant restriction sites in the population: the second and third sites on the map above.

c) individual 1: homozygous for **8 kb**

individual 2: homozygous for the 3+4 kb

individual 3: heterozygous for 7, 3+4

individual 4: heterozygous for 8,7

individual 5: heterozygous for 8, 4+3

individual 6: homozygous for 7

individual 7: homozygous for 4

d) heterozygous for 4, 7; 4, 8; 4, 3+4; 7, 4+3

e) Individual 2 could contribute only the 4+3 allele; individual 4 could contribute either the 8 or 7 kb. There are therefore 2 possible genotypes in the offspring: 8, 4+3; 7, 4+3.

f) Individual 2 got the 4+3 allele from each parent. Therefore the parents must be individuals 3 and 5, the only individuals having that allele of the RFLP.

9-7. a) Three alleles of the RFLP are present in this family: 13, 7.6, and 7.0 alleles. Individual II-1 must be homozygous for the Hb^S allele and inherited an allele from each parent. Each parent carried the 13 kb form and II-1 has this allele as well. Therefore, the 13 kb form of the RFLP is associated with the Hb^S form in this family.

b) To answer this question, look first at the affected individual II-2. He must carry two Hb^S alleles, but the RFLP alleles he carries are the 7.6 and 7.0. One of these alleles must have been associated with Hb^S in his mother and the other with Hb^S in his father. Individual I-2 is homozygous for the 7.6 kb allele but heterozygous for the Hb^S and Hb^A alleles. He must have passed on one copy of the 7.6 kb allele (and the one of the two that is associated with the Hb^S allele; the other chromosome with the 7.6 kb allele in this individual must have been associated with the Hb^A allele since I-2 is unaffected). Individual II-2 must have inherited the 7.0 kb allele from his mother, therefore the 7.0 kb allele is associated with Hb^S in I-1.

c) From the reasoning used in b), the 7.6 kb allele is associated with the Hb^S allele in individual I-2.

d) For an affected child to be born, both unaffected parents would have to be heterozygous. The probability is zero that a child of these parents would have sickle cell anemia because individual II-2 on the left does not carry the 13 kb form (associated with the Hb^S allele in that family), and therefore does not have the Hb^S allele.

e) The parents in the family on the left must be heterozygous to have had a child (II-1) who is homozygous. The probability that II-2 is heterozygous is 2/3. (Only three possibilities remain for an unaffected child of the heterozygous parents I-1 and I-2: 1/3 chance of homozygous; 2/3 for heterozygous.) The same rationale is true for II-1. If these two were heterozygous, there is a 1/4 chance of an affected child. Therefore the probability based on pedigree information alone is 2/3 × 2/3 × 1/4 or 1/9.

9-8. a) To determine the allele that is cosegregating with the disease, look at the affected individuals for a common allele, and at the unaffected for an allele that is not present. Allele 2 is found in all affected individuals except III-5 and is not present in unaffected individuals and is therefore linked to the disease allele in this family.

b) There must have been recombination between the RFLP locus and the disease locus in formation of the sperm in the father. The recombination results in allele 1 on the same chromosome as the disease allele. The recombinant sperm fertilized the egg and the zygote developed into individual III-5 in which allele 1 of the RFLP is now associated with the disease.

9-9. For an RFLP to be informative in a particular family, individuals should be heterozygous for a polymorphism. The two different alleles allow the inheritance of each chromosome (and therefore the inheritance of the RFLP and disease loci) to be traced from on generation to the next.

9-10. The parents must be heterozygous for the disease allele to have had an affected child. We can use the genotype of the affected individual to deduce which RFLP allele(s) in the parents are associated with the mutant CF allele. The probe is from the 5' end of a gene that is very large (250 kb) and the mutation could be in another region of the gene. Such a mutation would be very closely linked (>99%) but could be separated by recombination. The mother contributed the 5 kb allele to the affected child. Therefore the 5 kb allele in the mother corresponds to the mutant CF allele. The dad must have contributed the 3 kb allele, so this allele is associated with the CF mutation in him. The fetus inherited the 7 kb allele from the mom and the 5 kb allele from the dad and neither of these is associated with the CF gene mutation in the parents. The fetus is probably homozygous for the wild-type CF allele and will be unaffected (and is also not a carrier of the CF allele).

9-11. a) Sleeplessness is a Y-linked trait. The unusual feature of the hybridization is that the probe only hybridized with DNA from males. Since the Y is unique to males, this hybridization result suggest that the probe and therefore the gene is on the Y chromosome. The pedigree is consistent with this hypothesis since all affected individuals are male and we know that female is unaffected.

Individual II-3 would be XY^s (where s indicates the sleepless allele); II-2 and II-4 would be XX (no Y)

b) Two unaffected parents have an affected child which suggests the trait is autosomal recessive. But given the minimal information in the pedigree this trait could also be inherited as a sex-linked recessive. Look at the RFLP analysis. What alleles are present? Bands 2+4 are one allele, bands 1+3 represent the other allele. But in some individuals there is only one allele present and these are all male. This is explained by the probe detecting a sequence on the X chromosome. The trait is X-linked recessive. Genotype of II-1 is X^BY; II-3 is X^bY; II-4 is X^BX^b.

c) From the pedigree, inheritance of this trait could be any of the four possible modes of inheritance. Using the DNA data, there are two copies of each RFLP allele, so it is not X-linked. Affected individual I-1 has the same two RFLP alleles as does II-2 (affected) but individual II-4 (unaffected) has one of each allele. The disease allele cannot be dominant. The trait is therefore autosomal recessive. II-1 is *Ss*; II-2 is *ss*; II-4 is *Ss*.

9-12. The PCR analysis requires very little DNA, whereas RFLP analysis requires more DNA. PCR can be done using DNA from small tissue fragments and the DNA does not have to be well preserved. In addition, the PCR methodology is less expensive and time consuming because Southern blots and hybridizations are not required.

9-13. For in vitro fertilization the man would have to donate sperm, eggs would be collected from the female at ovulation and stored. After the fertilization had occurred in vitro and the embryos had progressed to the 8 cell stage, one of the cells would be removed, DNA prepared and the sample analyzed to determine the genotype of the embryo. PCR based analysis would be done on the DNA and compared with information on the parents and family members to determine the genotype of the embryo. If the genotype is suitable, the embryo would be introduced into the woman's uterus where it should implant to begin the pregnancy.

9-14. a) Look at the alleles present in this pedigree. The 4 + 1 kb is one allele; 2+1 is a second allele; 5 kb is a third. RFLP allele 4+1 segregates with the disease allele in the family on the left because affected individuals are homozygous for this RFLP allele and for the disease allele.
b) In the family on the right, the disease allele is associated with form 2 + 1.
c) Look first at the parents, II-4 and II-5 to understand the nature of the RFLP and disease alleles in III-1. Individual II-4 has the 5 kb allele from I-1 and the 4+1 (that is associated with the disease) from I-2, so is a carrier. Individual II-5 has the 2+1 (that is associated with the disease allele) from I-4 and the 4+1 (not associated with the disease allele) from I-3 . The child III-1 inherited the 4+1

allele associated with the disease from his mother (II-4) and the other 4+1, not associated with the disease, from his father (II-5). Because one of the RFLP alleles in III-1 is not associated with the disease allele in his family, the individual is unlikely to be affected.

d) Individuals II-4 and II-5 are both heterozygous. A carrier child would also be heterozygous but could have inherited the chromosome carrying the disease allele from mother II-4 (4+1 RFLP allele) or father (2+1 RFLP allele). The genotypes possible for a carrier are (4+1; 4+1) and (5; 2+1).

9-15. c), e), a), b), d)

9-16. a) At 100°C, no hybrids will form. (DNA would remain denatured as single strands.) At 80°C, the conditions are too permissive so that even oligonucleotide probes with a single base mismatch will hybridize. 90°C is a good temperature for differential hybridization using this allele specific oligonucleotide (ASO). Differentiation is made between DNAs having a complete match and those that have a single base different from the probe.

b) Individuals 1, 5, 6, 8 are homozygous for allele 1; individuals 2, 7, 9, 10 are heterozygous; individuals 3, 4 are homozygous for allele 2.

9-17. For ASO analysis, the probe should contain the base that varies and it is best to not have it on the very end. Therefore any 18mer that includes the base in bold would be a good probe. You need two different oligos for the analysis - one that contains a sequence corresponding to the mutant and the other corresponding to the wild type sequence.

9-18. The SSCP technique does not require any prior knowledge of nucleotide sequence of DNA and therefore is very useful when sequence and sequence variations have not been established.

9-19. If a closely linked RFLP is available and if the parents are heterozygous for the forms, the RFLP could be used to follow the inheritance of the mutant locus. Alternatively, polymorphism within the gene could be looked for using the SSCP technique. To do this, a region within the gene could be amplified by PCR, the DNA melted, and electrophoresed in its single-stranded state to look for distinctly different forms in the samples from each parent. If a variant is found and the parents are heterozygous, the linked sequence difference can be used to follow inheritance of the disease allele.

9-20. Minisatellites are usually too large to be amplified by PCR. Minisatellites are composed of repeats 10-40 bp in length and are present in tens to thousands of copies. The length of a minisatellite locus could be 100-40,000 bases in length. PCR is not usually used for amplification of fragments greater than a few thousand base pairs.

9-21. a) The child inherited the DNA represented by the second and third bands from the mom. The remaining one band corresponds to a band in the dad's DNA. Based on this analysis this man could therefore be the father but the result is not conclusive evidence of paternity. Additional polymorphisms could be examined to make a stronger case.

b) The two smallest fragments in this child's DNA correspond to two bands in the mom's DNA, but the other two bands do not correspond to the man's DNA. This man cannot be the father of this child.

9-22. a) To determine the probability that an individual in the Caucasian population would be expected to have these two alleles, the individual probabilities are multiplied (product rule of probability). The probability is therefore $(0.1)(.55) = .055$ or about 1/20 individuals would have this genotype.

b) In the Asian-American population, the probability is $(.08)(.01) = .0008$, or 8 in 10,000 (about 1/1000).

Chapter 10   The Mapping and Analysis of Genomes

Synopsis

This chapter describes the tools and goals of genome analysis, including the construction of physical and genetic maps that are ultimately used to locate genes and explore their function. Many of the molecular techniques you have learned in previous chapters are put to use in genome analysis. There are two main approaches used in genome analysis to associate a particular gene with its function. In the first approach, positional cloning, you begin with a phenotype and work to identify the gene(s) responsible for the phenotype. This approach is similar to the classical approach you studied in earlier chapters except in earlier analyses a mutation defined a gene as a factor in determining a phenotype, but the actual function of the gene product was harder to understand. In genome analysis, the goal is to get to the sequence of a gene through mapping and use that information to discover how the gene product works. In the second approach, you begin with a gene sequence and try to establish what phenotype the gene causes and therefore the role the gene plays. This bottom-up type of approach has increased in importance as more and more sequence data has been obtained for several organisms.

The characteristics and phenotypes geneticists are studying now are often far more complex than the traits that were studied in early experiments on peas and humans. Genome analysis has provides a wealth of data that allow us to ask more complex questions. Incomplete penetrance, phenocopies, and polygenic inheritance all provide challenges in establishing the relationship between gene(s) and phenotype.

Be prepared to:

After reading the chapter and thinking about the concepts, you should be able to:

- apply your knowledge of molecular techniques such as PCR and restriction analysis to solve genome analysis problems
- distinguish between genetic and physical maps and how they are made
- describe steps in positional cloning (from phenotype to gene clone)
- describe steps for understanding the function of a gene (phenotype it affects) starting with a cloned gene
- interpret data for locating genes in a DNA sequence
- know how to analyze genes involved in a complex quantitative trait

- distinguish between somatic and germ-cell therapy

Problem solving tips:

- Genomic libraries include all DNA in the genome; cDNA libraries include only expressed (transcribed) genes and lack intron sequences.
- Genetic maps are based on recombination frequencies.
- Remember there is usually trial and error involved in making restriction maps. Try to start with unambiguous, simple results to place a site and then progress to placing sites for an enzyme that cuts at multiple positions.
- The problems in this chapter are more representative of what is actually done in a molecular biology laboratory. Put yourself in the shoes of a researcher; be inquisitive and THINK!

Solutions to Problems

10-1. a) 9 b) 6 c) 1 d) 5 e) 7 f) 3 g) 4 h) 2 i) 8

10-2. The human genome contains a large amount of repeated DNA and intervening sequences (introns) that can be quite large. These two factors make the total DNA to protein-coding DNA ratio higher than that in bacteria.

10-3. a) The clones you isolated will contain microsatellite DNA sequences since that was the assay used to identify them. Do they contain unique sequences also? To determine this, you would have to cut the clone into small fragments, reclone the subfragments and use each as a probe in a hybridization with genomic DNA. If only a single band was seen after hybridization with one of these subfragments, you would have identified unique sequence within your clone.
b) To determine if the microsatellite associated with an STS is polymorphic in the population, the oligonucleotides used to detect the STSs surrounding a microsatellite should be used to amplify the DNA from a large sample of individuals in the population.

10-4. The FISH protocol for locating a gene on a chromosome involves a single hybridization with a labeled probe of your gene's sequence to a chromosome spread and therefore is very rapid. It is especially useful in organisms where the large numbers of matings required for linkage analysis are expensive (e.g., mouse) or not feasible (e.g., in humans) and the cytology is good (chromosomes are easily distinguished from one another). FISH can be used with any cloned piece of DNA, while polymorphic alleles are required for linkage analysis.

10-5. a) If the man is homozygous for one form or the other of the polymorphisms detected by the Alpha, Beta, or Delta primers, all of his sperm would have the same allele. Because the 200 bp fragment is never amplified, the man is probably homozygous for the sequences to which the Beta primers bind, preventing the fragment from being amplified. (This assumes that he does not have a deletion in this region of the genome). He is heterozygous for the sequences to which the Alpha and Delta primers bind because the sequence is amplified in some of the sperm but not in others.

b) A map can be generated for the sequences recognized by the Alpha and Delta primers. If they are linked, an allele of one will cosegregate with a particular allele of the other. The amplified forms (400 and 100 bp) are cosegregating in 10 cases; the unamplified forms cosegregate in 11 cases; either one is seen alone in 4 cases. The sperm cells in which only one of the amplified alleles is seen are the result of recombination between the two loci. 4/25 or 16% is therefore the recombination frequency, so the loci recognized by the Alpha and Delta primers are 16 map units apart. We do not know anything about the position of Beta with respect to Alpha and Delta because the man was homozygous for the Beta locus and the segregation of the two copies of the gene with respect to the other alleles could not be followed.

10-6. a) Linkage distances expressed in centimorgans are based on recombination frequencies. The difference in genetic distances between the sexes can be explained by a difference in the amount of recombination overall in the genomes of males and females.

b) The physical differences between males and females would not have to differ based on the map distance differences calculated from recombination. In fact, the physical differences are the same, as is required to maintain the integrity of the chromosomes passed from generation to generation. (e.g., grandmother to father to daughter).

10-7. a) The question asks for information on sites in the genomic insert, but since your data is given based on the plasmid, make a map for the sites on the plasmid, but indicate which sites are in the vector and which are found in genomic DNA. Because the NotI and SalI sites are so close to and flank the BamHI site where the fragment inserted, place these on either side of your vector sequence. (The small fragments between the NotI, BamHI and SalI sites can be ignored.) The vector alone is 2.5 kb in length. The fragments (4.2 kb NotI, 3.0 kb SalI, 2.5 kb NotI + SalI) containing vector sequences are indicated by the hybridization. Looking at the SalI digest, the 6.8 kb SalI site is not cut by NotI since it is still present in the double digest (NotI + SalI). The 3.0 kb SalI fragment contains the vector. The 2.5 kb vector plus the 0.5 kb in the NotI + SalI digest adds up to the 3.0 kb SalI fragment while the 2.3 + 1.7 NotI + SalI fragments add up to the 4.0 kb SalI fragment. Place the SalI site 0.5 kb from the NotI site. Try the 6.8 kb SalI adjacent to the 0.5 kb NotI-SalI fragment. The second NotI site cannot be in the 6.8 kb SalI fragment, so the NotI fragment extends from the NotI site on the one side of the vector, through the first SalI site and the 6.8 kb SalI fragment and the 2.3 kb beyond that site (0.5 + 6.8 + 2.3 = 9.6 kb NotI fragment). From the map you can see the SalI sites that are present in the genomic region. (Remember that the NotI and SalI sites next to the ends of the vector are not present in the chromosome.)

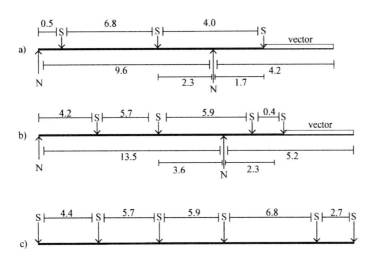

b) The hybridization in this case indicates the overlap of DNA sequences between clone 1 and 2 (and will include those fragments containing vector sequences as well). Notice that only the 2.3 kb NotI + SalI genomic fragment seen in clone 1 is also present in clone 2. Map the sites in clone 2 in a similar way to how it was done for clone 1. Of the SalI fragments, the 5.7 and 0.4 kb fragments are not cut by NotI because they are still present in the NotI + SalI digest. The other fragments, 6.7 and 5.9, are cut into 2.5 + 4.2; 3.6 + 2.3 kb fragments, respectively, in the double digest. The 2.5 kb NotI + SalI fragment is the vector fragment, so the SalI fragment containing the vector must be the 6.7 kb fragment and the SalI site can be placed 4.2 kb beyond the NotI vector site (at left end of the figure below). The NotI site has to be located 13.5 kb away from the NotI site at the end of the vector to accomodate the placement of the SalI site and the fragments present in the double digest. What NotI-SalI fragments can add up to this fragment? We know the 4.2 kb NotI + SalI fragment is part of this NotI fragment; that leaves 9.3 kb. If the 5.7 kb SalI fragment were within this 13.5 kb fragment, that would leave an additional 3.6 kb NotI + SalI fragment. But how do we know the order of the 5.7 and 3.6 kb fragments? Looking at the hybridization, the 3.6 kb fragment hybridizes with the clone 1 probe but the 5.7 does not. Clone 1 and 2 overlap in the 2.3 kb region and parts of adjacent fragments. The 3.6 kb should therefore be next to the 2.3 kb NotI-SalI fragment (which also helps us place the next restriction fragments- 2.3 and 0.4 kb on our map).

c) From the maps of clones 1 and 2 we can put together an overlapping map of the SalI sites. We know from previous hybridizations that there is overlap through the 2.3 kb SalI-NotI region, and we know that the 2.3 NotI-SalI fragment is part of a 5.9 SalI fragment. The fragments flanking this are 5.7 (from clone 2) and 6.8 (from clone 1). The fragment next to the 5.7 has to be the 4.4 kb fragment (the only one not already accounted for in the clone 2 hybridization to genomic DNA). Next to the 6.8 kb fragment is the 2.7 kb fragment (the only fragment not accounted for in the clone 1 hybridization to genomic DNA.)

10-8.   The four cosmids are aligned below. Notice that the third cosmid had to be flipped over for the alignment.

10-9. The two cosmids probably contain DNA sequences that are present as repeat DNAs on many different chromosomes including the one from which the inserts in all the other cosmids came from. Some section of the probe DNA is this repeated sequence and therefore cosmids from other parts of the genome hybridized to it.

10-10. In positional cloning of a human gene, a correspondence is made between the inheritance of the disease allele and a particular region of a chromosome using affected and unaffected individuals within several families and molecular markers that have already been mapped. This process starts at a low level of resolution (sometimes by identifying the chromosome on which the disease gene is found) and progresses to a finer resolution. Once the disease locus has been mapped to within 1 cM of a marker, you can shift to examining physical clones containing DNA in this region (1 cM = approximately 750 kb). If a set of clones already exists that cover the region of the genome you identified, you can search for genes within those clones.

10-11. One method is to search for transcribed regions in the cloned segment using Northern hybridization (DNA probe versus RNAs from the cells). A second method is to search the DNA sequence for open reading frames (with codon usage appropriate for the species) and for sequences that are splice sites (indicating the presence of introns). A third method is to hybridize the DNA with DNA of a related mammalian species such as mouse. Cross-hybridizing regions suggest that the region was maintained without major changes and therefore could contain a coding region.

10-12. The best way to prove that a candidate gene identified by positional cloning is in fact the gene in which mutations cause a disease is to look for alterations in the gene sequence that are consistently found in affected individuals but not in unaffected individuals. Another type of evidence is expression in tissues predicted by the disease. Such evidence is suggestive but not conclusive.

10-13. For all the transcripts shown, we don't know the direction in which transcription occurs and the primary transcript could extend beyond the 3' end of the gene. The 4.1 kb mRNA was produced by a primary transcript that extends through all the DNA between A and G because bands A, G as well as bands D, F in between hybridized to the transcripts. There must be introns in the regions covered by probes B,C and E. The 3.4 kb mRNA was produced by a primary transcript that extended from fragment B into fragment E. This could be two or more introns since B and C are adjacent fragments. There is at least one intron in the region covered by fragment D. The 1.8 kb transcript was produced by a primary transcript that extends from fragments D into G. Because the mRNA is only 1.8 kb but the total fragment length in this region is 3.5, there must be at least one intron in this region, but we cannot say in which fragment(s) from this data.

10-14. a) The size of the cDNA1 plasmid is 6.93 kb. You know this either by the HindIII digest in which only one linear fragment was generated or by adding the sizes of fragments from another digest. The insert is therefore 6.93 - 4.2 (vector size) = 2.73 kb.
b) The pGEN1 plasmid is 7.73 kb. The genomic insert is therefore 7.73 - 4.2 = 3.53kb.
c) For the cDNA clone: The 4.6 kb PstI fragment must extend from the PstI site in the vector through all of the vector and into the insert. Place the PstI sites on a map of the clone including the vector restriction sites. This placement shows you where the 0.9 kb BamHI-PstI fragment comes from. The largest BamHI (4.4 kb) fragment is the only one large enough to include the vector. In the BamHI-PstI digest, the 3.7 kb fragment is within the vector (see the vector map). The BamHI-PstI 1.52 kb fragment must be adjacent to the PstI site at the end of the 0.9 kb BamHI-PstI fragment to lead to a 2.42 kb fragment when the plasmid is digested with BamHI only. The 0.11 BamHI fragment does not change in size when digested with any other enzyme. This information allows you to place the site giving rise to this fragment. All that remains in the insert is 0.5 kb. Together with the 0.2 kb from

the adjacent vector, that makes up the 0.7 kb BamHI-PstI fragment. HindIII cuts the 2.42 kb BamHI fragment into 1.32 and 1.1. To accommodate the HindIII-PstI digestions, the HindIII site 1.32 kb from the BamHI site has to be 0.42 from the PstI site. HindIII cuts the 2.33 PstI fragment into a 1.91 and 0.42 kb fragment, consistent with the map as drawn.

For the genomic clone: Notice that there is an additional HindIII site in this clone and the BamHI and PstI fragments have increased by 0.8kb. The additional DNA must be in the BamHI 3.22 and PstI 3.13 kb fragments. In the HindIII-BamHI digest, the 1.1 kb increases to 1.7 kb and a new 0.2 kb fragment appears. See map below for the location of the additional HindIII site and the map of this clone.

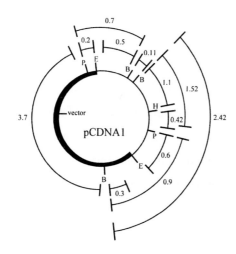

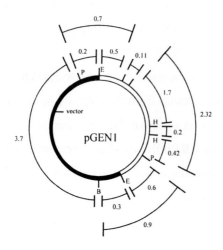

d) There is a minimum of one new intron of 0.8 kb size because there is an additional 0.8 kb of DNA in the genomic clone.

e) The DNA sequence could have been searched to locate splice donor and acceptor sites.

10-15. cDNA clones are made by synthesizing a DNA copy of a processed mRNA. The two different cDNA clones can be explained by different splicing of the same primary RNA. Two transcripts that differ in content can be produced by splicing different information from the primary message. The DNA produced by reverse transcription of the two messages will therefore have different content.

10-16. Because cDNAs are made from mRNAs, they lack regulatory region information and intervening sequences (intron DNA).

10-17. If pseudogenes arose from cDNAs, they would lack the introns found in the functional copy of the gene.

10-18. a) The two hybridization signals come from the transgene inserted at a site on one chromosome and the wild-type gene at its normal location.

b) Two copies of the transgene inserted into the chromosome. The third signal is from the wild-type copy again.

c) You might expect a stronger hybridization signal with the three copies. (In fact, it is difficult to distinguish a single copy from several copies using FISH.)

d) With two cuts within the transgene, an internal fragment is generated. If three copies are present, the intensity of three copies of the internal fragment compared with the one copy of the portion of the transgene that forms a joint fragment with the adjacent chromosomal DNA would be easy to detect.

10-19. a) Once the DNA sequence had been obtained, the amino acid sequence of the protein product can be deduced. Tell-tale features of protein sequences are motifs that have been shown in other proteins as having a specific function.

b) Analysis of the tissues in which the gene is expressed (transcribed) provide useful clues about function. Knocking out a gene and looking for a phenotype caused by the knock-out is another type of analysis that might lead to information on the gene.

10-20. a) The presence of a zinc-finger motif (DNA sequence that encodes the amino acids capable of forming the zinc finger in a protein) is suggestive that the protein functions as a transcription factor.

b) Similarity throughout the new sequences with a previously identified gene would suggest that the two genes had a common ancestor and could be part of a gene family.

10-21. The mRNAs present in all tissues could be alternately spliced messages and therefore the proteins made could differ in some regions. The antibody for the brain protein may not be able to recognize such proteins that are made from the alternate messages. Another possibility is that the two other messages are not translated in any tissues.

10-22. a) The genome of yeast is very small (there is very little repetitive DNA) compared with other eukaryotes and therefore the cloning of genes is easier. Any gene involved in basic cell functions would be a good gene to clone in yeast.

b) The genome of *C. elegans* is small relative to humans but worms have several functions that are similar to, yet more primitive than humans, such as a nervous system. Some of the basic functions of nerve cells can be studied this model system.

c) The mouse is the closest model organism to humans and genes that are homologous to disease genes in humans can be cloned and analyzed.

10-23. a) The gene size is often large due to multiple introns (which are sometimes large in size). Cloning a cDNA can get around this problem.

b) To get expression in only certain tissues, the gene being introduced should be cloned next to a regulatory region that leads to expression in the tissues.

10-24. Germ-line therapy results in changes in the genome that are passed on to future generations. With random integration, the transgene could insert into and disrupt some other gene that could lead to another disease.

10-25. Identifying subcategories of the same disease and doing the genetic analysis on the subcategories may help identify the genetic basis of the disease. For example, early and late onset of Alzheimer disease have been shown to be dependent on predisposing alleles in different genes.

10-26. When analyzing quantitative traits, organisms with the extreme phenotypes are crossed. For kernel size in wheat, plants with the largest kernels (probably all homozygous for alleles of all genes involved) would be crossed with plants having the smallest kernels (probably homozygous for all the alternate alleles of the genes involved).

Chapter 11   The Eukaryotic Chromosome: An Organelle for Packaging and Managing DNA

Synopsis

This chapter describes the structure of eukaryotic chromosomes and how that structure affects function. The very long, linear DNA molecules are compacted with proteins in the chromosomes to fit into the nucleus. Several structures are essential for duplication, segregation, and stability. Replication origins are necessary for copying DNA during S phase; centromeres are necessary for attachment to spindle fibers and proper segregation during cell division; telomeres are necessary at the ends of the linear DNAs to maintain the integrity of the DNA molecule. Chromatin structure (packaging of DNA in the chromosomes) can have consequences for gene activity. Areas of normally packaged chromosome can become decompacted for expression to occur. Some regions of chromosomes or entire chromosomes are packaged in a different way that decreases gene activity (for example, heterochromatin or Barr bodies).

Be prepared to:

After reading the chapter and thinking about the concepts, you should be able to:

- describe the essential elements of eukaryotic chromosomes

- predict the stability of artificially constructed chromosomes based on the components they contain

- analyze data on changes in chromatin compaction

- predict phenotypes or genotypes of individuals in a lineage in which inheritance of an imprinted gene is being followed

Problem-solving tips:

- Origins are necessary for replication; centromeres are necessary for proper segregation during mitosis and meiosis.

- For maternally imprinted genes, the copy of the gene inherited from the mother will not be expressed in either her sons or daughters. For paternally imprinted genes, the copy of the gene inherited from the father will not be expressed in either his sons or daughters.

- A mutant of a gene that is imprinted must be inherited from the non-imprinting parent to see expression of the mutant phenotype.

- Put yourself in the position of being the researcher. When designing experiments consider the aim of the experiment, the concepts that apply to the problem, and think through experimental methods you know to find a relevant methodology.

Solutions to Problems

11-1. a.4 b.10 c.5 d.8 e.9 f.2 g.3 h.6 i.1 j.7

11-2. You could make a probe using the cloned DNA and hybridize the probe to a blot made from the pulse-field gel. Since all chromosomes are separated, you would be able to match the hybridization signal with the position of a chromosome on the original gel.

11-3. During metaphase, chromosomes are more compacted than interphase chromosomes. In interphase, the chromosomes are compacted 40-fold and during metaphase the chromosomes are compacted 10,000 fold.

11-4. Core histones H2A, H2B, H3, H4 make up the core octamer around which DNA is wound. H1 is found outside the nucleosome core and is involved in the next level of compaction, formation of a 300 Å fiber.

11-5. Non-histone proteins associated with chromosomes include those involved in structure and packaging of chromosomes (e.g., scaffold proteins); replication; segregation (e.g. kinetochore); transcriptional regulation and processing; recombination.

11-6. a) To replicate the longest chromosome (66Mb) from a bidirectional origin of replication, 33 Mb would have to be copied during the 8 minute cycle (=480 sec). If a single origin of replication was used and replication took the entire 8 minutes of the cycle, the rate of polymerization per second would be 0.06875 mb/sec or 68.75 kb per second.
b) If origins of replication are 7 kb apart and are bidirectional, 3.5kb would have to be replicated during every 8 min division cycle. The polymerization rate would be 3.5 kb/480 sec or 7.3 bases per second.

11-7. a) To examine the end of one specific chromosome, your probe DNA would have to contain unique DNA found next to the repeated TTAGGG sequences.

b) The bluriness of the band indicates that the hybridizing fragments from one end of the chromosome in a population of cells are not homogeneous in length. In other words, the fragment at the end of the chromosome in all cells are not the same size. The number of repeat sequences at the telomere (and therefore the telomere length) varies from cell to cell, especially in actively dividing cells.

11-8. The new sequences that are added must be specific for the species which is adding them (in this case, yeast). Because the YAC was transformed into yeast, telomerase in the yeast cell added on sequences specific for yeast.

11-9. a) A plasmid containing only the URA^+ gene would have to integrate into the chromosome to be replicated and maintained because is has no origin of replication. Once integrated it will be stably maintained.

b) A URA^+, ARS plasmid could be maintained separately as a plasmid or integrate into the chromosome. If it remains as a plasmid, it would not be very stable mitotically and would be lost from many of the daughter cells. If integrated, it would be very stable.

c) The URA^+, ARS, CEN plasmid could only be maintained as a separate plasmid in the cell. (If it did integrate into the chromosome, there would be two centromeres on that chromosome and during mitosis the chromosome would break.) The plasmid would be very stable from one generation to the next because the centromere sequence directs its segregation.

11-10. a) You could use the cloned yeast gene as a probe to hybridize to clones in a human cDNA library. (Note: To identify related genes in distantly related species, the stringency of the hybridization conditions is often lessened so you are not demanding that every base be identical.)

b) You could make an antibody directed against the human protein as the antigen. In fixed cells, the antibody would bind specifically to this protein and using a tag on the antibody (usually fluorescence), the location of the protein in the cell (nucleus vs cytoplasm, for example) could be determined.

c) There are several possibilities here. If the protein acts in chromosome segregation, a temperature-sensitive mutation in the gene encoding the protein at high temperature could lead to loss of chromosomes- aneuploidy. If homologous chromosomes had genetic markers that allowed them to be distinguished (heterozygous for several genes on the same chromosome), the phenotype of aneuploidy could be observed. Or the loss of function might disrupt chromosome segregation in a way that is recognized as a signal to stop the cell cycle until the error is corrected. (Such signals are described more fully in Chapter 17). Or, if the protein is part of the kinetochore structure, at high temperature there might be no kinetochore formed and chromosomes would not migrate to the daughter cell. The result would be a polyploid mother cell.

11-11. The fragments that show a high percentage of Trp$^+$ colonies after 20 generations without selection for the plasmid are the fragments that contain the centromere DNA. These are the 5.5 kb BamHI, 2.0 kb BamHI-HindIII, and 0.6 Sau3A. Because the smallest of these has high mitotic stability and its ends are within the boundaries of the other fragments, the centromere sequence must be contained within that fragment.

11-12. You could digest the BAC, the YAC, and the genomic DNA with several restriction enzymes and compare the restriction patterns of each. To see the pattern of fragments in the normal chromosome, the genomic DNA would have to be hybridized with a probe containing the BAC or YAC DNA.

11-13. Boundaries prevent the decomposition of chromatin from spreading. The opening up or decomposition of chromatin leads to expression of genes in that region. Removal of a boundary could cause more genes in regions adjacent to the boundary to be expressed.

11-14. b. The DNA in cell type II is very accessible to DNAase. This is characteristic of an open chromatin configuration that is found near a highly expressed gene.

11-15. a) Centromeric DNA is constitutive in *Drosophila* and the region around a gene such as white is facultative.
b) Centromeric DNA is also constitutive in humans. Inactive X chromosomes (Barr bodies) are an example of facultative heterochromatin.

11-16. a) In the presence of a *Su(var)* mutations, there would be fewer white patches in the eye and more red patches compared to a homozygous *Su(var$^+$)* fly. The situation would be the opposite (more white patches and fewer red patches) if the fly were homozygous for the *E(var)* mutation.
b) The *Su(var)* and *E(var)* mutations have a phenotype that would lead you to think the proteins encoded by the genes are involved in chromatin condensation. Assuming the mutations are loss of

function (null) alleles, this would imply that *Su(var⁺)* genes encode proteins that establish and assist spreading of heterochromatin, because loss of some of the gene product results in engulfment of neighboring genes by heterochromatin. The *E(var)⁺* genes could encode proteins that restrict the spreading of heterochromatin, since loss of one copy of the gene allows heterochromatin to spread into neighboring genes more often. The results also suggest that position effect variegation is very sensitive to amounts of either type of protein because a reduction of 50% of either type of protein causes the effect of the phenotype.

11-17. a. 1 b. 0 c.1 d. 1 e. 3 f. 0

11-18. The twin sisters could be monozygotic (came from one fertilization event) but the X that was inactivated in cells affected by muscular dystrophy are different. Both girls would be heterozygous for the DMD gene, but only one was affected because of the random inactivation of X chromosomes.

11-19. Girls with a genotype of $X^{CB}X^{cb}$ could have some patches of cells in the eye in which the X chromosome carrying the CB allele was inactivated and therefore those patches would be defective in color vision. Usually, enough cells have the X^{cb} allele inactivated (and therefore X^{CB} is active) and is sufficient color vision and therefore no phenotypic effect of the X^{cb} cells.

11-20. Polytene chromosomes are large and therefore easily visible in the light microscope. The banding pattern is reproducible from one fly to the next, providing a map of the chromosomes based on the landmark bands.

11-21. The copy of the gene inherited from the father is inactive when there is paternal imprinting. An affected son (having a mutant allele that causes the phenotype) must have inherited the active but mutant allele from his mother.

11-22. a) T b) T c) F d) F

11-23. a) *Aa* b)*Aa* c) *Aa*

11-24. a) Because the *IGF-2R* gene is maternally imprinted, the allele(s) expressed by the children were inherited from the father. Since one child expresses the 50K and the other expresses the 60K protein, the father must have been heterozygous. From this limited information, we know only one allele that the mother carries- the 60K allele.

b) We know that Bill Jr. inherited the 60K protein from his father, because the copy from his mother was imprinted and therefore not expressed. Because one of his children expresses the 50K protein, we know he was heterozygous for 60K/50K. The 50K protein was not expressed in Bill Jr. and therefore was the maternally imprinted allele he received from his mother. We can now say that Bill Jr.'s mother was also heterozygous 60K/50K.

Chapter 12   Chromosomal Rearrangements and Changes in Chromosome Number Reshape Eukaryotic Genomes

Synopsis

Rearrangements of sections of chromosomes by duplication, insertion, deletion, inversion, or translocation can affect distances between genes and the function of genes in which they occur. Very large chromosomal rearrangements can be seen microscopically as changes in banding patterns. Many rearrangements are detectable by changes in linkage or effects on meiotic products. Crossing-over during meiosis within an inverted region in an inversion heterozygote leads to imbalanced gametes, so it appears from resulting viable gametes that recombination is suppressed within that region. Transposable elements are segments of DNA that can move from one position to another in the genome. Different types of elements move using a transposase enzyme or by reverse transcription of RNA into a DNA copy.

Changes in chromosome number can be the result of loss or gain of one chromosome (aneuploidy) or changes in the numbers of sets of chromosomes (e.g., polyploidy). Aneuploid cells are generally inviable in humans, with the exception of those that involve sex chromosomes (where there are still phenotypic consequences of extra or lost chromosomes).

Be prepared to:

After reading the chapter and thinking about the concepts, you should be able to:

- trace products of a crossover within an inverted region of an inversion heterozygote when the inversion is either paracentric or pericentric
- use deletion data to map a gene
- design experiments to determine if there are inversions, deletions, or translocations in a strain
- predict gametes produced by a translocation heterozygote

Problem solving tips:

- Deletion of a gene on one homolog (deletion heterozygote) uncovers a mutation in the gene on the other homolog.
- Deletions, inversions, and translocations change the linkage of genes that surround or are within the rearrangement.
- Deletions of DNA can be analyzed using restriction analysis. In a diploid organism, a deletion on one chromosome will mean that a restriction fragment that comes from within the deleted region

of the genome will be at half the concentration of that found in a normal cell that has the DNA on both chromosomes.

- Draw out chromosomes of parents and progeny from complicated multigenerational crosses so you are very clear about the chromosome composition going into and out of meiosis. This should help you predict the genotypes and phenotypes of progeny.

- A set of chromosomes refers to the collection of chromosomes as found in a gamete. For example, in humans, a set of chromosomes is the 22 autosomes + one sex chromosome.

Solutions to Problems

12-1. a) 4 b) 8 c) 6 d) 5 e) 7 f) 3 g) 2 h) 1

12-2. a) Because each of the strains is mutant for one or more of the marker genes and two of the spores die, X-rays induced deletion mutations.

b) Two spores die in each ascus because the deletions removed some essential genes from one of the chromosomes in the diploid.

c) Because there is only one X-ray induced mutation per strain, if more than one gene is defective in a strain, those genes must be deleted on the same chromosome. Using this type of logic, you can deduce that the four genes, w, x, y, and z, are on the same chromosome.

d) The order is $w\ y\ z\ x$. Genes w and y are deleted in strain 1, uncovering the w^- and y^- mutations, so w and y must be adjacent. Genes x,y,z are deleted in strain 2 so must be adjacent; therefore w must be to one side of x and z , the order of which is unspecified by this information. From strain 3 we can determine the order of x and z. Strain 3 is deleted for w, y and z, therefore the gene that follows y must be z.

12-3. In polytene chromosomes, characteristic banding patterns are seen in each chromosome. If there is a duplication of a region of DNA, the bands in that region should be duplicated on one homolog. Therefore an extra copy of the bands would be present in the looped out region. (A total of three copies of that banding pattern would be present in the preparation.) If the mutation is a deletion, the looped out region in a heterozygote would contain the only copy (wild-type) homolog of all those bands.

12-4. *javelin*- A6; *henna*- between C2-3 and D2-3.

The deletion data allows you to narrow down the region in which the gene lies. Because deletion A results in expression of the mutant phenotypes javelin and henna, at least part of these genes must lie between A2-3 and D2-3. Deletion B, with the phenotype henna, indicates that part of *henna* lies between C2-3 and E4-F1. Because this deletion strain is javelin$^+$, it also indicates that all of *javelin* must be before C2-3. The C and D deletions being wild-type tells us that all of the *henna* gene is found before D2-3. The finding that inversion B is wild-type indicates that neither gene is encoded in B4. Inversion A has a mutant javelin phenotype, indicating that javelin is encoded by DNA in A6. Very few *Drosophila* genes extend beyond one band, so can probably assume A6 is the location of javelin. If we don't make this assumption, we would conclude that *javelin* is between A2-3 and B4 because inversion B indicates that the gene cannot extend beyond 65B4. The left end cannot be delimited by the data given.

12-5. a) 3.0   6.3   4.2   5.6   0.9

Fragments that are deleted in one chromosome will be lighter in intensity than those that are not deleted (and are therefore present in two copies per cell). Looking at the fragments that are deleted, you can tell which fragments are contiguous and by analyzing all the deletions you can tell the order of the fragments. New bands that appear in only one of the deletion strains represent new joint fragments generated by the deletion and therefore are not useful for this analysis. For each of the strains, the following deleted fragments will be considered:

Strain	complete fragments deleted
Strain 1	6.3, 5.6, 4.2
Strain 2	6.3, 4.2, 3.0
Strain 3	5.6, 0.9
Strain 4	6.3, 3.0

Looking at strains 1 and 4, the 6.3 kb fragment is deleted in both strains. The other bands must represent fragments that lie to either side of the 6.3 kb fragment. From strain 4, we know that the 3.0 kb fragment lies to one side of the 6.3 fragment, but since it is not lost in strain 1, the 5.6 and 4.2 kb fragments must lie to the other side (although we don't know the order yet). Looking at strain 2 that also has a deletion of the 6.3 kb fragment, the deletion also includes the 3.0 and 4.2 kb fragment. This tells us that the 4.2 kb fragment must be immediately adjacent to the 6.3 kb fragment and the order is 3.0, 6.3, 4.2, 5.6. Strain 3, in which the 5.6 and 0.9 kb fragments are deleted, indicates that the 0.9 is adjacent to the 5.6 kb fragment but since it was not deleted in strain 1 it must be on the side opposite the 4.2 kb fragment.

b)

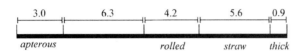

Based on the phenotypes of each of the heterozygous strains, the approximate location of genes can be determined. Using the restriction map derived from part a) and the phenotype of the strains, *rolled eyes* and *straw bristles* look like they are found somewhere within the 6.3, 5.6, 4.2 kb fragments. But since Strain 3 also has the straw bristles phenotype, that gene can be placed at least partly within the 5.6 kb fragment. Strain 2 has the rolled eyes phenotype also, so the gene must be in one or both of the fragments that are common to strain 1 and 2 (6.3 and 4.2 kb). However, since Strain 4 has a deletion of fragment 6.3 but is not mutant for *rolled eyes*, the gene must lie within the 4.2 kb fragment. *Apterous wings* is mutant in strains 2 and 4 which have in common deletion of the 6.3 and 3.0, but since the mutant phenotype is not seen in Strain 1, the gene lies in 3.0 kb fragment. *Thick legs* is mutant in strain 3 that is deleted for the 5.6 and 0.9 kb fragments, but is not mutant in strain 1 that is deleted for the 5.6 kb fragment, so the gene lies in the 0.9 kb fragment.

12-6. a) yellow, zeste, split   and   tan, zeste, normal bristles
The majority of the progeny result from no crossing-over in the females. The females have the genotype

$$\frac{y \qquad z^{1} \qquad w^{+R} \qquad spl}{y^{+} \qquad z^{1} \qquad w^{+R} \qquad spl^{+}}$$

The male progeny inherit their X from their mothers and will inherit either one of the two chromosomes shown above unless crossing-over occurs.

b) Classes A and B are due to crossing over anywhere between the *y* and *spl* genes resulting in the reciprocal classes:

$$y \qquad z^{1} \qquad w^{+R} \qquad spl^{+}, \quad \text{and} \quad y^{+} \qquad z^{1} \qquad w^{+R} \qquad spl$$

which correspond to yellow, zeste, normal bristles and tan, zeste, split bristles respectively.

c) Classes C and D are due to unequal crossing over between the two copies of the w gene. (w^{+R} is shown here as two copies of the w^+ allele). The misalignment can occur in two different ways but each one produces one class of chromosomes that result in a wild-type eye color.

Misalignment 1

$$y \qquad z^I \qquad w^+ \; w^+ \quad spl$$
$$\underline{\qquad\qquad\qquad X \qquad\qquad}$$
$$y^+ \qquad z^I \qquad\quad w^+ \; w^+ \qquad sp^{l+}$$

Misalignment 2

$$y \qquad\quad z^I \qquad w^+ \; w^+ \qquad spl$$
$$\underline{\qquad\qquad\qquad X \qquad\quad}$$
$$y^+ \qquad z^I \qquad w^+ \; w^+ \; spl^+$$

The results of crossing-over that lead to one copy of w (which leads to the wild-type eyes) are $y^+ \, z^I$ $w^+ \, spl$ and $y \quad z^I \, w^+ \, spl^+$. (The reciprocal recombinant for each of these misalignments would have three copies, so would have the zeste phenotype and would be indistinguishable from flies in classes A and B.)

12-7. The diploid cell contains a pericentric inversion in one homolog. The pairing of the four chromatids present in meiosis in the inversion heterozygote during meiosis I are shown below. Use this drawing to trace the consequences of crossovers in different regions.

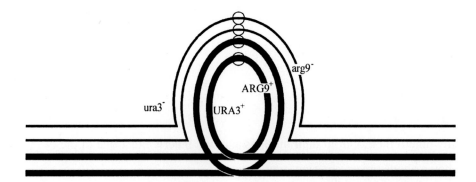

a) A single crossover outside the inversion area will produce four viable spores in a tetrad: two $URA3^+ \; ARG9^+$ spores and two $ura3^- \; arg9^-$ spores.

b) A single crossover between $URA3$ and the centromere results in a duplication of one region on one of the resulting chromosomes and a loss of that region on the other chromosome. This genetic imbalance is usually lethal, so the two spores containing the products of the recombination will die.

The two chromatids that are not involved in the recombination will be viable have the phenotype Ura$^+$ Arg$^+$ and Ura$^-$ Arg$^-$.

c) The double crossover produces four viable spores, two of which are Ura$^+$ Arg$^+$ and two Ura$^-$ Arg$^-$.

12-8. In this problem, the diploid cell contains a paracentric inversion on one homolog. The pairing of the chromosomes in the inversion heterozygote are shown below. Use this drawing to trace the consequences of crossovers in different regions.

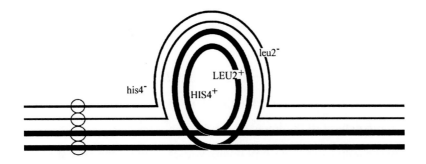

a) A single crossover within the inverted region (between *HIS4* and *LEU2*) leads to a chromosome containing two centromeres (dicentric) and a chromosome lacking a centromere (acentric). Neither type of chromosome segregates properly so the spores that receive these chromosomes will definitely die. The two remaining chromatids that are not involved in the crossover will be present in two viable spores, His$^+$ Leu$^+$ and His$^-$ Leu$^-$.

b) In a double crossover with both crossovers occurring between the *HIS4* and *LEU2* genes, all four spores are viable, with 2 His$^+$ Leu$^+$ and 2 His$^-$ Leu$^-$.

12-9. a) 2, 4. Inversion loops are seen during pairing only if the cells are heterozygous for an inversion.

b) 2, 4. Genetically imbalanced chromosomes (having deletions and duplications of regions of the homologs, respectively) are generated by crossovers in a pericentric inversion (4). Dicentric and acentric chromosomes (also genetically imbalanced) result from crossing-over within the paracentric inversion (2).

c) 2. An acentric (and the reciprocal dicentric) fragment is produced from a single crossover within a paracentric inversion in an inversion heterozygote.

d) 1,3. If the inversion is found on both homologs (cell is homozygous for the inversion), crossovers within the inversion will not affect the viability of the spores.

12-10. a) The order in X-ray: *a*   *b*   *c*   *f*   *e*   *d*   *g*   *h.*

The order in Zorro: *a*   *b*   *f*   *e*   *d*   *c*   *g*   *h*

You are told that Bravo, X-ray, and Zorro all contain the same number of bands. This information, together with the altered recombination frequencies between genes, indicates that there must be inversions in the strains. In X-ray, frequencies are altered in the *c-d, d-e, e-f, f-g* interval. The inversion must therefore occur between *c-d* and *f-g*. The *c-d* and *f-g* intervals must still include some non-inverted DNA to see crossing-over that produces viable gametes. Similarly, the genetic distance is lowered in the *b-c* and *f-g* intervals for Zorro, where the inversion end points are found and minimal in those intervals completely within the inversion.

b) The physical distance in the X-ray homozygotes between *c* and *d* is greater than that found in the original Bravo homozygotes. The inversion occurred in this portion of the chromosome, so *c* and *d* are now separated by many more genes (all the inverted DNA).

c) The physical distance between *d* and *e* in the X-ray homozygotes is the same as that found in the Bravo homozygotes because this interval is completely within the inverted segment. The relationship of *d* to *e* has not changed.

12-11. a) If *yg* is on an chromosome that was not involved in the translocation, it will behave as an independent genetic element from the translocation. The semisterile F_1 is a translocation heterozygote and would be expected to produce 1/2 fertile, 1/2 semisterile progeny from an alternate segregation. (Products of adjacent segregation are inviable.) For the leaf color phenotype, 1/2 of the progeny are wild-type (green); 1/2 are yellow green. Using the product law of probability, 1/4 will be fertile and green, 1/4 fertile and yellow-green, 1/4 semisterile and green, 1/4 semisterile and yellow-green.

b) If the translocation involved chromosome 9, the fate of the fertility and leaf color phenotypes are connected

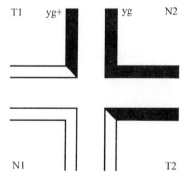

From the translocation heterozygote, the products of an alternate segregation are balanced and viable. The progeny will be 1/2 fertile and yellow-green (N1 + N2) and 1/2 semi-sterile and green (T1 + T2).

c) The rare fertile, green and semisterile, yellow-green result from crossing-over between the translocation chromosome and homologous region on the normal chromosome with which it is paired. The N1+ N2 chromatids, after crossing-over, will contain yg^+ and the T1 + T2 will contain yg^-

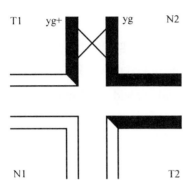

The distance between the translocation breakpoint and the *yg* gene could be determined using the frequency of crossing-over (represented by the rare fertile green and semisterile yellow-green progeny).

12-12. a) 0 white, dead spores. The inversion has no effect if crossing-over does not occur.

b) 4 white spores out of 8 total spores. Only two chromatids are involved in the crossover. The products of crossing over between the two chromatids are unbalanced gametes that would probably die. The remaining two chromatids survive as meiotic products and divide mitotically to form 4 viable spores in the ascus.

c) 0 white spores. If a crossover occurs outside the inversion loop when a paracentric inversion is present on one chromosome, all products are viable.

d) 8 white spores. All of the resulting gametes would be genetically imbalanced and would die.

e) 0 white spores. Alternate segregation produces balanced gametes.

f) 0 white spores. The crossover in the translocated region would simply reciprocally exchange DNA between homologous portions of the chromosome. All spores live.

12-13. a) 1, 3, 5, 6

b) 2, 4

c) 1, 3 (arise from alternate segregation pattern)

d) 5, 6 (arise from adjacent-1 and -2 segregation patterns)

12-14. a) Four combinations of the *cn* and *st* alleles are possible in equal proportions (1/4 of each) from the heterozygous male by independent assortment:

       cn   *st*   white eyes

       cn   *st+*   cinnabar eyes

       cn+   *st+*   wild-type eyes

       cn+   *st*   scarlet eyes

When these gametes fertilize a white-eyed female, 1/4 of the progeny will be white-eyed.

b) The *cn* and *st* or cn^+ st^+ allele combinations seem to be linked now. This could occur if there had been had a translocation between chromosome 2 and 3 with *cn* and *st* ending up on the same chromosome. The normal chromosomes (N1 + N2) come from the mother and the cn^+, sn^+ are linked in the translocation chromosomes (T1 + T2).

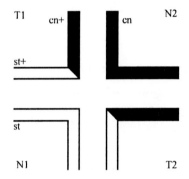

c) The wild-type F_1 females would contain the balanced translocation containing the cn^+ and st^+ alleles on one of the translocation chromosomes. The pairing in meiosis would be the same as shown for the male in part b) except that now, in the female, crossovers could occur. A crossover between *cn* and the translocation breakpoint followed by an alternate segregation would produce gametes containing N1 + N2 balanced chromosomes with the genotype cn^- st^+ (phenotype cinnabar) and balanced translocation T1 + T2 chromosomes with the genotype cn^-st^+ (phenotype scarlet).

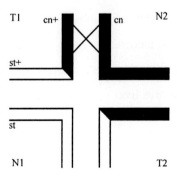

12-15

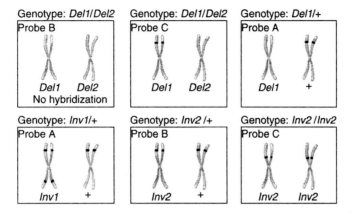

12-16. Insects that are homozygous for the translocations, when they mate, will produce progeny that are translocation heterozygotes. The fertility of these progeny should be reduced by about 50% and half of their progeny will also have reduced fertility.

12-17. The progeny will be only *Lyra* males and wild-type (*Lyra$^+$*) females. Remember that the Y chromosome pairs with the X chromosome during meiosis, so the N1 + N2 chromosomes will be the autosome on which *Lyra* is normally found and the X chromosome.

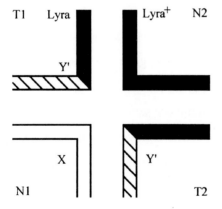

There are only two kinds of genetically balanced gametes produced by the male by alternate segregation. T1 and T2 would yield a gamete with both elements of the translocation between the Y and the autosome, producing Lyra males when fertilized with X. N1 and N2 would yield a gamete with the X chromosome and the wild-type *Lyra$^+$* autosome; this would produce wild-type females (*Lyra$^+$*) when fertilized with *Lyra$^+$* gametes from the wild-type mother.

12-18. Transposase acting on the ends of transposons that are near each other on the chromosome but have normal chromosomal DNA containing genes between them could transpose the entire large segment of DNA.

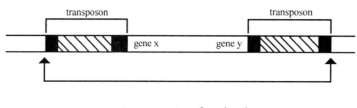

transposase acts on far ends and
transposes the whole section of
the chromosome

12-19. *Ds* is a defective transposable element unable to encode its own transposase, while *Ac* would be a complete, autonomous copy of the same transposon. Chromosomal breakage at *Ds* insertion sites would be a possible side product of the transposition mechanism catalyzed by the *Ac* transposase. Insertion of *Ds* into a gene might yield a mutant allele that is unstable in the presence of *Ac* transposase. *Ac* is likely to be complete and contain ends that can be recognized by the transposase, because *Ac* can transpose itself to different chromosomal locations (explaining the different locations of Ac in different strains).

12-20. The stronger *ct* mutant alleles could result from imprecise excision of the *gypsy* element, leading to the deletion or alteration of sequences within the *ct* gene that would strongly affect its functions. Alternatively, these could result from the movement of the *gypsy* transposon into other parts of the gene that would compromise gene function more seriously. The stable ct^+ alleles are likely to be precise excisions in which the *gypsy* element has moved out of the gene, restoring the normal ct^+ sequence. The unstable ct^+ alleles are likely to be cases in which the transposition process altered the *gypsy* element in the *ct* gene so that the gene could function normally. However, the bit of the transposon that remains may try to transpose and these attempts might cause deletions or rearrangements of the *ct* gene.

12-21. You could use a probe DNA composed of the sequence preceding the 200 A residues and do a hybridization to genomic DNA. You would expect to see several hybridizing bands if other copies of a retroposon are present.

12-22. It is easiest to work this type of problem if you figure out the sizes of the genomic fragments and place these on the map.

A) 5. The base change exactly at coordinate 6.8 will alter the restriction recognition site and therefore the 1.1 and 3.0 kb fragments will not be seen as separate fragments. One new 4.1 kb fragment is now present.

B) 3. A point mutation at 6.9 has no effect on these restriction digests since an EcoRI site is not found at that position.

C) 2. This deletion will remove DNA within the 2.5 kb fragment, resulting in a new fragment of size 2.2 kb.

D) 6. This deletion removes DNA found in two restriction fragments (1.1 and 3.0 kb). Because the deletion also removes the recognition site, one new fragment (3.8 kb) will be seen.

E) 8. The insertion of a transposable element at coordinate 6.2 will change the size of the 1.1 kb fragment so the 1.1 kb fragment is no longer seen and a new fragment is present.

F) 10. An inversion with breakpoints at 2.2 and 9.9 will alter the two fragments in which the endpoints are located - the 5.7 kb and 2.5 kb fragments but will not affect the 1.1 and 3.0 kb fragments.

G) 7. The reciprocal translocation will affect the 2.5 kb fragment in which the breakpoint occurs and two new fragments will be seen that contain the part of the 2.5 kb that remains in the original chromosome and the DNA that is part of a different chromosome,

H) 1. A reciprocal translocation with a breakpoint at 2.4 will affect the 5.7 kb only. The 5.7 kb fragment is no longer a digestion product.

I) 4. An additional 2.0 kb added within the 3.0 kb fragment, creating a 5.0 kb fragment.

J) 9. The 2.5 kb fragment is maintained but the 4.6 increases to 6.6 kb. Because there is a restriction site within the region that is tandemly duplicated, a new fragment of 2.0 kb is now present.

12-23. a) The *x* number in *Avena* is 7.

b) Sand oats (14) are diploid. Slender wild oats are tetraploid. Cultivated wild oats are hexaploid.

c) Sand oats: 7; slender wild oats: 14; cultivated wild oats: 21

d) The *n* number for each species is the number of chromosomes in the gametes and therefore is the same as the answer in c.

12-24. a) 15 b) 13 c) 21 d) 28

12-25. Possibility A would be due to non-disjunction during meiosis II in the father. B would be due to non-disjunction in meiosis I in the mother. C could arise by non-disjunction in meiosis I in the father. D could arise by non-disjunction in the mother in meiosis II.

12-26. The somatic mosaic Turner individuals could have arisen from a mitotic nondisjunction event in the early divisions of the embryo or from chromosome loss. The latter might be more likely because mitotic non-disjunction in a normal XX embryo should produce an XXX daughter cell in addition to an XO, while mitotic non-disjunction in an XY embryo yields an XO and an XYY daughter cell. The XXX and XYY cells were not reported in the karyotype analysis, although it is possible that XXX or XYY daughter cells simply did not expand into large clones of cells during development.

12-27. Haploid plant cells in culture can be treated with colchicine to block mitosis (segregation of chromosomes) during cell division and create a daughter cell having the diploid content of chromosomes. Once the diploid resistant cell is obtained, it could be grown into an embryoid. Proper hormonal treatments of the embryoid will yield a diploid plant.

12-28. What are the gametes the $F^A F^a F^B F^b$ plant can produce? Because the chromosomes of same ancestral origin still pair, the cross can be represented as a dihybrid cross between heterozygotes (treating the resistance genes from the two ancestral species as independently segregating genes). Recall that 9/16 will have the F^A-F^B- genotypes and these will be resistant to all three pathogens.

12-29. In karyotype analysis, chromosomes are stained and have a characteristic banding pattern. The banding patterns of the homologs in the autopolyploids should be the same, but the chromosomes of the different species that formed the allopolyploids would probably have different banding patterns.

Chapter 13   The Prokaryotic Chromosome: Genetic Analysis in Bacteria

Synopsis

This chapter describes characteristics of genetic analysis in bacteria, with a focus on gene transfer in *E. coli*. Two key features have made *E. coli* a powerful model organism for understanding basic cell processes. First, the ability to grow massive numbers of bacteria on defined media allows easy and quick isolation of mutants. Second, there are many naturally occurring ways to transfer DNA from one cell to another. The DNA transfer mechanisms - transformation, conjugation, and transduction (general and specialized) and the ways in which geneticists use these are described. In addition to their value for research, DNA transfer mechanisms are important for survival and evolution of bacterial species.

Be prepared to:

After reading the chapter and thinking about the concepts, you should be able to:

- distinguish between selection (only one specific genotype can grow) and screening (more than one genotype can grow, but additional analysis is needed to establish the genotype of each cell.)
- describe how Hfr and F'cells are formed and the uses of each for mapping genes and complementation analyses
- set up Hfr crosses to map genes (describe the genotypes of donor and recipient and the selective media used)
- analyze time of entry data to map genes
- analyze time of entry data to map origins of transfer for Hfr strains
- map genes using cotransduction or cotransformation frequencies
- describe the differences between transformation, transduction, and conjugation

Problem solving tips:

- The F plasmid can integrate at different locations around the *E. coli* chromosome to generate Hfr strains that have different origins of transfer.
- Genes closest to the origin of transfer are more likely to be transferred into a recipient than those further from the origin.
- The time at which genes are transferred into a recipient in an Hfr cross is a reflection of the distance from the origin of transfer.

- Hfr crosses are usually used to get a low resolution map; P1 transductions are useful for finer genetic mapping.

- Bacteriophage lambda can integrate into the chromosome (lysogeny) to generate a lysogen.

- F plasmids and bacteriophage lambda can excise imprecisely, picking up adjacent chromosomal genes.

Solutions to problems

13-1. a) 4 b) 5 c) 2 d) 7 e) 6 f) 3 g) 1

13-2. From the second to last tube, 200 colonies should grow. The total dilution that had been made was 10^{-4} so the concentration of cells was now 2×10^3/ml or 2000 cells/ml. 0.1 ml was spread so 2000 cells/ml x 0.1 ml or 200 cells were spread on the plate. The last tube had 10-fold fewer bacteria. The culture had been diluted 10^{-5}, so there were now 2×10^2 cells/ml or 200 cells/ml. The 0.1ml that was spread therefore contained 20 colonies.

13-3. a) iv. With lactose as the carbon source, the cells would have to be able to use lactose as a carbon source to grow (Lac^+). Lac^- cells would not be able to grow, so this is a selection for Lac^+ cells.

b) iii. The rich media allows both Lac^+ and Lac^- to grow, so this is not a selective media, but the Xgal distinguishes between the two phenotypes.

c) ii. To select for Met^+ cells, the media should be lacking in methionine, demanding that the bacteria be able to synthesize methionine to grow.

13-4. To determine the number of nucleotides you would need to know to identify a gene, you need to figure out how many bases would represent a unique sequence in a DNA molecule of the *E. coli* genome size (5Mb or 5,000,000 bases). Since there are 4 bases possible at each position in a sequence, 4^n represents the number of possible combinations that can be made with n number of bases. For example, 4^2 is the number of unique sequence combinations that could be made with 2 positions. Therefore, if you were looking for a unique 2 nucleotide sequence, you might expect to find it, on average, in 16 nucleotides. A sequence of 11 nucleotides would appear $1/4^{11}$ or one in 4×10^6 bases (4 Mb); a sequence of 12 nucleotides would appear uniquely $1/4^{12}$ or one in 16.8 Mb. You would therefore need to sequence about 12 nucleotides in order to know they came from a unique place in the *E. coli* genome. This problem could also be solved setting up and using the equation $4^n =$

5×10^6 (5 Mb). Solving for n, you would rearrange the equation to read: $n\log 4 = \log 5 \times 10^6$. Solving for n: = 11.1, so you would need more than 11 nucleotides to find a unique nucleotide sequence.

b) If you had a 4 amino acid sequence, this would correspond to 12 nucleotides (which you determined in part a) as being sufficient to find a unique sequence), but because of degeneracy of the genetic code, you would in reality know the identity of about 8 of these nucleotides. (For the amino acids that have 6 codons, the you would potentially know one less nucleotide.) If you had a sequence of six amino acids, you would probably know at least 12 unique nucleotides that encode that amino acid. (Statistically, there are a few sequences of 6 amino acids that appear in more than one protein, so a few more amino acids would be even better.)

13-5. You could isolate genomic DNA from *E. coli B* and *E. coli K*, digest with EcoRI, electrophorese the DNA, transfer to a filter for a Southern hybridization using an IS1 DNA as a probe. The number of bands that appear will correspond to the number of IS1s in the genome because EcoRI does not cut within the IS element. (If a band is twice as intense as other bands, this would indicate there are two fragments of the same size that contain an IS1 and the band represents two IS1 elements.)

13-6. You could do a mating between the mutant cell with 3-4 copies of F and a wild-type F-recipient. If the mutation were in the F plasmid, you would expect a recipient strain into which the plasmid was transferred would have the higher copy number. If the mutation was in a chromosomal gene, the higher copy number phenotype would not be present in the recipient after a transfer to a wild-type strain.

13-7. You could test to see if the plasmid encoded the toxin genes by transferring the plasmid (by transformation) into a non-toxin producing recipient strain. If the gene was encoded on the plasmid, the transformant would now produce the toxin.

13-8. The *purE* and *pepN* genes would be cotransformed at a lower frequency if the pathogenic strain was used as a host donor strain. There is a lower likelihood that the two genes would be on the same piece of DNA because they are separated by more DNA than in the non-pathogenic strain.

13-9. Selecting for Pyr$^+$ exconjugants is selecting for an early marker. The frequency with which genes beyond *pyrE* are transferred decreases with distance from this early marker. This experiment then examines time of entry of markers. The order of genes is:

  pyrE xyl mal arg met tyr his

13-10. a) The Hfr strain requires cysteine. It would grow on minimal glucose media containing this amino acid. The F$^-$ strain requires tryptophan, histidine, tyrosine, and threonine and therefore would need these amino acids in the media to grow.

b) In the first experiment, the matings were not disrupted as they were in experiment 2, so some mating pairs continued to transfer DNA even after the cells were diluted and plated. To get exconjugants, both cysteine and/or trp must have been transferred, so at 8' when the mating was interrupted in experiment 2, both of these genes had not yet been transferred.

c) Man$^+$ exconjugants are selected by plating on minimal medium containing mannose instead of glucose and supplemented with cysteine, tryptophan, histidine, tyrosine, threonine, streptomycin.

d.)

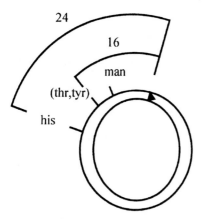

e) Fewer recombinants will be formed when quadruple crossovers are necessary to generate the wild-type. You can use this to determine the order. For the second cross 4, in which fewer wild-type colonies arose, arrange the genes in an order that is consistent with quadruple crossovers.

  man$^+$ tyr$^-$ thr$^+$
  ———— X ———— X ————
  man$^-$ tyr$^+$ thr$^-$

The order is *man tyr thr*.

13-11. The order is *lac arg102 arg101 arg103*. To get an Arg$^+$ phenotype, a recombination has to occur that results in the introduction of the wild-type sequence in one part of the *arg* gene from the donor but maintenance of the wild-type region of another part of the *arg* gene. Depending on the order of mutations, either double or quadruple crossovers are needed to reconstitute the *arg$^+$* gene. For example, in cross A, *arg102$^+$* must be transferred into the recipient but the *arg101$^+$* allele must also be maintained to get the Arg$^+$ phenotype. Because this occurs at a relatively high frequency (.50%), the order of the genes must be *lac, arg102 arg101*. In contrast, in cross B, *arg101$^+$* allele must be transferred in and the *arg102$^+$* allele maintained in the recipient. For this to occur, there must be 4 crossovers (based on the order of genes that we just determined), and therefore the frequency is of occurrence is much lower (.02%). (Note that all the frequencies are low because there is only a small region in which crossover can occur.) A similar logic is used to determine that *arg 101* must be between *lac* and *arg103*.

13-12. The order is *ilv bgl mtl*. This is a three factor cross in which the selected marker, *ilv*, is the farthest from the origin. The major class of exconjugants will be the one in which all three genes were transferred from the donor (Ilv$^+$ Mtl$^+$ Bgl$^+$). The smallest class is the one in which quadruple crossovers had to occurred. The double crossover class tells us what the order is. In this case, the Ilv$^+$ Bgl$^-$ Mtl$^+$ is the smallest class so *bgl* must be between *ilv* and *mtl*. (The Ilv$^+$ Mtl$^-$ Bgl$^-$ class results from a crossover between *ilv* and *mtl* and the Ilv$^+$ Mtl$^-$ Bgl$^+$ class comes from a crossover between *mtl* and *bgl*.)

13-13. a) The F plasmid must have integrated into the chromosome near the *mal* genes and then excised incorrectly, picking up adjacent *mal* genes.

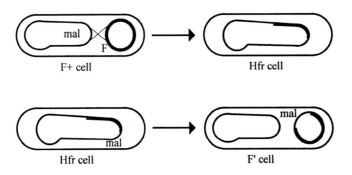

b) The F'mal must have recombined with the chromosome to generate an Hfr strain.

13-14. a) *pab ilv met arg nic (trp pyr cys) his lys*

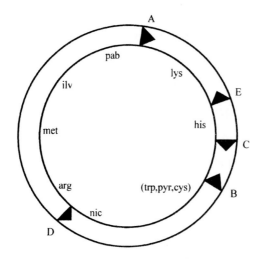

b) The class of recombinants that has the smallest number is the class that requires four crossovers instead of two. When Trp⁺ was selected, most of the exconjugants will have also recombined in the adjacent *pyr*⁺ *cys*⁺ alleles from the donor (seen here as the largest class containing 790 recombinants.) There is a small class (5 recombinants) represented by Trp⁺ Pyr⁻ Cys⁻.These are the result of four crossovers and is produced only if *trp* is in the middle (*cys trp pyr*). The two other classes represent double crossovers between *cys* and *trp* or *pyr* and *trp*. To find the distance between the genes divide the number of events in which recombination occurred between two genes (include single and double crossovers) by the total number of recombinants scored. Between *cys* and *trp*, there were (145+5)/(790+145+60+5) = 150/1000 = 15%. Between *pyr* and *trp*, there were (60+5)/1000 = 6.5%.

c) F' plasmids are formed by inappropriate excision of an F plasmid from the chromosome. To isolate an F' carrying the *trp, pyr, cys* genes, the strain from which the F' is derived should have the F plasmid integrated next to these genes. HfrB or HfrC would therefore both be candidate strains for the isolation of the F'.

13-15. From Hfr #1, the genes scored are in the order

     M   Z   X   W   C

From Hfr #2,

     L   A   N   C   W

From Hfr #3

     A   L   B   R   U

From Hfr #4

      Z   M   U   R   B

Combining this data requires looking for the overlapping map orders and presenting the order of genes in the same direction around a circular map. Then use the differences in time of entry for adjacent genes to get distances.

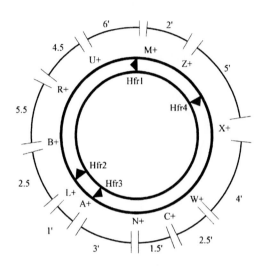

b) To isolate an F', you need to start with an Hfr strain that has the F plasmid integrated near the genes you would like to isolate on an F' plasmid. The easiest way to identify the F' you want is to start with an Hfr strain that transfers the genes you are interested in late in the mating. In this case you would use Hfr3 to isolate an F' carrying the N$^+$ gene. Then if you screen for a derivative of that Hfr strain that transfers the marker early, the genes should be on an F' rather than being transferred by the Hfr. The screen could be done by patching many individual isolates of the Hfr strain and doing a replica mating to a plate containing a N$^-$ recipient and select for N$^+$ cells. (Another alternative is to screen for transfer of N$^+$ in a *recA* mutant. To get an N$^+$ exconjugant cell from an Hfr requires *recA*-mediated recombination, but no recombination is required for transfer and maintenance of an F' plasmid in the recipient cell.)

13-16. The assay will detect *recA* mutants because recombination is required in the recipient F$^-$ to obtain the exconjugants. The screen would be for a *recA* mutation in the recipient cell which is were the recombination occurs.

13-17. Transduction is a type of gene transfer mediated by phage. The DNA is protected inside the protein head of the phage. Transformation is gene transfer using naked DNA (not enclosed in any protective structure). The DNA being transferred by transformation will therefore be susceptible to degradation by DNase. If the transfer still occurs after DNase treatment, transduction must be occurring.

13-18. a) The time of transfer for four of the markers *(gly, phe, tyr, ura)* is indistinguishable, so we cannot put them in an order on the map but *lys, nic* and the cluster of genes can be placed.

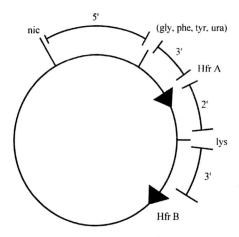

b) *Phe* was cotransduced with *ura* more frequently than with *tyr*, so the order is *phe-ura-tyr*. (None of the cotransduction classes is very rare, so none of these result from quadruple crossovers). The relationship of these three genes to other markers and the order of *gly* gene is still unknown.
c) To map the *gly* gene with respect to other markers, you could select for Gly⁺ transductants and score other markers to determine which genes are cotransduced with the highest frequency with *gly*. The Gly⁺ transductants would be selected on minimal+lysine+phenylalanine+tyrosine+uracil, then the transductants would be tested for growth on media lacking one or more of the unselected amino acids.

13-19. In generalized transduction, chromosomal DNA is packaged randomly into transducing phage particles that contain chromosomal DNA only, no phage DNA. In specialized transduction, the phage in the lysate contains chromosomal DNA covalently attached to phage DNA. The region of the chromosome DNA that is transduced is the same in all transducing particles in the lysate.

13-20. a) The lysate will contain 10^2 Gal$^+$ specialized transducing phage and there will be 100 Gal$^+$ colonies obtained.

b) 1. Infect the new Gal$^-$ mutants with lambda to obtain a lysogen. (Lysogens can be recognized by the inability of bacteriophage lambda to reinfect the cell.) 2. Induce the lysogen. (The lysate should contain some specialized transducing phages now carrying the *gal$^-$* mutation.)

3. Infect each of the nonlysogenic tester strains with the lysates from each mutant. Isolate a lysogen.

4. The phenotype of the lysogen will indicate whether the genes complemented or not. (Note you have to create a lysogen for this to work. There could also be a recombination between *gal* genes also to generate the Gal$^+$ cells, and this would not tell you if there was complementation and what gene the mutation was in.)

c) Mutant 1 is a *galC$^-$* mutant. Mutant 2 is a *galB$^-$* mutant. Mutant 4 is a *galB$^-$* mutant.

d) deletion (This could also be a regulatory mutant, see chapter 15)

e) There are two copies of the galactose genes. One set of genes is the original chromosomal DNA and the other is on the lambda transducing phage that integrated at the lambda attachment site right next to the *gal* genes.

f) The single colonies could arise from recombination event between homologous DNA in the integrated lambda transducing phage copy and chromosomal copy of the *gal* genes. Recombination could result in a single wild-type copy of the *gal* genes left on the chromosome.

13-21. a) The newly isolated mutation would be in the F$^-$ strain and the Hfr strains would be wild-type. The basic plan is to do enough matings to cover the whole genome, select for exconjugants and test those exconjugants to see which Hfrs have donated the wild-type copy of the gene at a high level. Tetracycline resistance (encoded by the Tn10) would be the selectable marker from the Hfr and the F$^-$ should have a counterselectable marker (Str^r). To cover the genome, you would need to do 20 matings (There are 100 minutes of chromosome and an Hfr strain with origins every 5 minutes are available). The experiment could be done as described here. Set up each mating by mixing strains. The matings could be done as patch replica platings. Grow up a plate spread with recipient so it produces a confluent lawn of cells and grow up a plate containing patches of the different Hfr strain. Replica plate each of these onto the same velvet and print that onto the selection plate (tetracycline + streptomycin). Check the selected exconjugants for the mutant phenotype by replica plating from the selection plate onto two rich plates- one incubated at 30° (permissive temperature for the mutation) and the other at 42° (non-permissive temperature).

b) The effective average plasmid size is 20 kb - (2 × 1 kb for overlap on each end) or 18 kb. The genome is 4.6 Mb in size, so 4600 kb/18kb = 255 plasmids (and therefore 255 transformations).

13-22. 1. First pass an extract from *E. coli* cells (in which β-galactosidase is present) through an APTG-agarose resin. β-galactosidase would bind to the resin along with any proteins that in turn bind to β-galactosidase. (You can then remove these proteins from the column using a detergent.) 2. Digest the mixture of proteins with trypsin to generate a large number of smaller peptides. 3. Subject this mixture to mass spectrometry to get the molecular weight of the peptide fragments. The molecular weights of each of the peptide fragments will be unique because of the range of molecular weights for individual amino acids. 4. Take this profile of peptide molecular weights and compare it in the computer to a list of all the expected molecular weights for all triptic fragments of all proteins in *E. coli*. (This can be determined from the genome information.) This should tell you which proteins are present in the mixture, and therefore the genes that encode proteins binding to β-galactosidase.

b) You can use the APTG-agarose resin for the general methodology if you make a fusion between your protein coding gene and the *lacZ* gene. The fusion gene cloned into a plasmid will produce the fusion protein in a cell. An extract from the cell containing the clone can be passed over an APTG-agarose column and proceed as in part a). You would get out at the end molecular weights of tryptic fragments of proteins that bind to your protein as well as those that bind β-galactosidase (whose identity you already determined in part a).

Chapter 14   The Chromosomes of Organelles Outside the Cell Nucleus Exhibit Non-Mendelian Patterns of Inheritance

Synopsis

Mitochondria and chloroplasts are organelles that contain their own genomes. The genomes of both share similarities with prokaryotic genomes- in the types of genes, organization of genes, as well as chromosome structure. The presence of several organelles per cell in most organisms leads to many different combinations of organelle types in the same cell. Heteroplasmy refers to the presence of more than one type of genome per cell – homoplasmy is the state in which there is uniformity of the genome copies. Inheritance of organelle genomes is not dependent on the same cell machinery as nuclear chromosomes use for mitosis and meiosis. In most organisms, organelles are inherited from one generation to the next from the mother. This leads to distinct patterns of inheritance as seen in pedigrees.

Be prepared to:

After reading the chapter and thinking about the concepts, you should be able to:

- differentiate between heteroplasmy and homoplasmy experimentally
- recognize organelle inheritance in human pedigrees
- suggest explanations for unusual phenomenon involving organelle genomes that may arise from heteroplasmy and different proportions of affected cells

Problem-solving tips:

- Characteristics of mitochondrial gene inheritance in humans are that children of affected mothers are affected but children of affected fathers are never affected.
- There can be several organelles in a cell and several copies of the genome in each organelle.

Solutions to problems

14-1. a) 6 b) 8 c) 7 d) 2 e) 1 f) 3 g) 5 h) 9 i) 4

14-2. a) both b) mitochondria c) chloroplast d) chloroplast e) mitochondria

14-3. a) both b) both c) neither d) both

14-4. You would want to be sure that the nuclear and chloroplast DNAs separated and purified by density differences are pure enough so that there is not significant cross-contamination of DNAs. To do this, you should use a probe from a gene known to be nuclear encoded in plants and red algae and also a probe from a gene known to be encoded in the chloroplast of both organisms. If the nuclear DNA sample from the red alga was significantly contaminated with chloroplast DNAs, a positive hybridization signal might not indicate that the gene is nuclear encoded, but the hybridization could have been due to contaminating chloroplast DNA in the nuclear DNA preparation.

14-5. a) The universal code is used for nuclear encoded genes

UGG	CAU	AUA	AUG
	CAC		
Trp	His	Ile	Met

b) There are variations in the universal code that are found in mitochondria

UGG	CAU	AUC	AUG
UGA	CAC	AUU	AUA
Trp	His	Ile	Met

14-6. a) The large subunit of Rubisco is encoded in the nuclear genome of all three species, but the small subunit is nuclear encoded in the green alga and chloroplast encoded in the red and brown alga.
b) Because the size of the transcript for the red and brown alga is the same for both the large and small subunit probe, they appear to be cotranscribed. This is consistent with the fact that both genes are chloroplast encoded. The transcripts for the green algal Rubisco subunits are different sizes and therefore represent different transcripts. The genes are not cotranscribed as would have to be true for genes located in different genomes.

14-7. a, c, and e are characteristics that are bacterial-like. Choice b) is not related to bacterial genomes, since the universal code is used in those organisms. Choice d) is not related to bacterial genomes, since introns are not a universal feature of bacterial genomes.

14-8. The best way to determine if the entire sequence of the gene is present is to clone the gene from the nuclear DNA and determine the DNA sequence that region of DNA

14-9. With repeated backcrossing, the nuclear genome becomes more similar to the original male fertile nuclear genotype. The fact that sterility still occurred with each cross indicates that sterility was not a nuclear gene effect.

14-10. a). If you have probes that will differentiate between the two different genomes and have a recognizably different tag on the probe (different fluorescent tags, for example), you could do a double hybridization to cells in situ and see if both were hybridizing to the same organelles. PCR amplification of DNA from a population of cells would not indicate the state of individual organelles. For example, there could be cells containing all a^+ mitochondria and other cells in the population containing all a^- mitochondria. When DNA from such a population is used for PCR amplification, both mitochondrial genomes would be represented, suggesting heteroplasmy although individual cells are homoplasmic.

14-11. a) 3 b) 1 c) 2

14-12. Mitochrondrial DNA from one parent can be excluded from entering the zygote or can be destroyed once in the zygote.

14-13. Offspring would resemble the mother because the sperm does not contribute significantly to the cytoplasm of the zygote.

14-14. Homoplasmic cells would have difficulty surviving because of the loss of energy metabolism (in the case of mitochondria) or of photosynthetic capability (in the case of chloroplasts).

14-15. a) The diploid would be resistant to erythromycin and would have the genotype ery^r. The ery^s version of the mitochondria would have been destroyed.
b) The genotype would be heterozygous, ery^r/ery^s and the phenotype would reflect the dominant allele.
c) The diploid could be put under sporulation conditions and if the gene were nuclear, 2 spores would be ery^s and two would be ery^r. If the gene were mitochondrial, all four spores would be ery^r.

14-16. Mate the C^r MATa strain with a C^s MATa strain and sporulate the diploid. If the gene is nuclear, the chloramphenicol phenotype would segregate 2:2. If the gene is mitochondrial, there will be 4:0 segregation.

14-17. a) 2 b) 1 c) 4 d) 3

14-18. All of the offspring of an affected female are affected. None of the affected males offspring are affected (unless the mother is affected also).

14-19. a) One possibility is that the mother (I-1) may have had mitochondria in which only a small proportion of the genomes were mutant, while the proportion of mutant genomes in the daughter was much higher. An alternate explanation is that the mutation occurred in the germline of individual I-2 so she wasn't affected but her offspring II-2 received the defective mitochondria.
b) You could look at the mitochondrial DNA from somatic cells in the mother. If the mutation occurred in her germline and was passed on to II-2, her somatic cells would not show any defective DNA.

14-20. 1) c 2) a 3) b

14-21. The variation in tissues affected could be due to differences in the time during development when the mutation occurred. In two individuals, a mutation may have occurred in cells that give rise to different sets of tissues. Variation in which tissues are affected and also in severity of disease could also be due to the proportion of mutant genomes in the cells of each tissue.

14-22. Gel electrophoresis is better suited for an overview because deletions can be very large and might not be amplified. In addition, sequences to which primers bind might be deleted in some mutations so no information could be obtained about the size of the deletions.

Chapter 15   Gene Regulation in Prokaryotes

Synopsis

This chapter describes gene regulation in bacteria including the genetic analysis that led up to the postulation of the operon theory - the paradigm of gene regulation. A basic principle derived from experiments on the *lac* operon is that proteins bind to DNA to regulate transcription. How mutations in the components of the regulatory system proved Jacob and Monod's theory is an instructive lesson in the power of the genetic approach for understanding basic cellular processes. With the development of molecular biology techniques and increased study of protein structure, the operon theory of gene regulation as been refined so we understand more about DNA binding proteins and their interactions with regulatory regions. Experiments on the lactose operon lead to the development of fusion technology in which the *lacZ* gene is placed next to a regulatory region of another gene. Expression of that gene could then be monitored by measuring expression of β-galactosidase. Another type of fusion was developed in which the regulatory region of the *lac* operon was fused to a gene whose expression was then controlled using the induction of the *lac* genes.

In addition to the negative and positive control described for the *lac* operon, global transcriptional regulation based on changes in RNA polymerase and its subunits is described as is attenuation- a mechanism for fine-tuning transcription.

Be prepared to:

After reading the chapter and thinking about the concepts, you should be able to:

- predict overall cellular expression of proteins when there are site or gene mutations on a chromosome and other mutations on the F plasmid (merodiploid analysis)
- distinguish between positive and negative regulators based on the behavior of mutants
- propose models for regulation of a set of genes based on mutant and molecular analyses
- describe the process of attenuation and how mutations in the component parts might affect expression of the *trp* genes or other amino acid operons that are regulated by attenuation
- design molecular experiments to test predictions of models based on mutational analysis
- describe use of and distinguish between *lacZ* fusions (geneX-*lacZ*) and lac regulatory fusions (*lac*-geneX)

Problem solving tips:

- Proteins diffuse in the cytoplasm and therefore can act on any copy of their binding site in a cell.

- Regulatory sites in the DNA only affect DNA adjacent to them.

- Sites and proteins that bind to sites form the control system of gene regulation.

- Regulatory proteins that bind to several molecules (e.g., DNA, another protein, inducers) have distinct regions or domains in the protein for these interactions.

Solutions to Problems

15-1. a) 4 b) 8 c) 5 d) 2 e) 7 f) 1 g) 3 h) 6

15-2. The *rho* gene must be essential. Only conditional mutations can be isolated for essential genes.

15-3. The non-lysogenic recipient cell did not have the cI (repressor) protein in the cytoplasm, so the incoming infecting phage could go into the lytic cycle. When phage infects a lysogen, the repressor is present in the cytoplasm so can bind and repress the incoming phage.

15-4. A mutation in a site affects only adjacent DNA so a promoter mutation would act in cis.

15-5.	β-galactosidase *(lacZ)*	permease *(lacY)*
a)	constitutive	constitutive
b)	constitutive	inducible
c)	inducible	inducible
d)	no expression	constitutive
e)	no expression	no expression

15-6. These mutations could be changes in the base sequence of the operator site that compensate for the mutation in *lacI*. For example, if an amino acid necessary for recognition of operator DNA is changed in the *lacI* mutant, a compensating mutation changing a recognition base in the operator could now cause the mutant LacI protein to recognize and bind to the operator.

15-7.

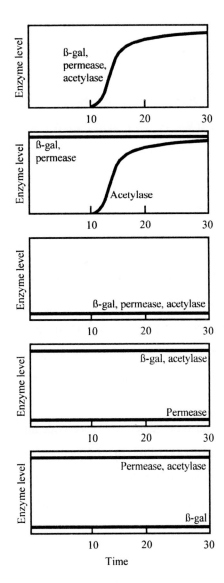

15-8. a) The operon is negatively regulated because a defect in the protein binding site leads to constitutive synthesis. When the negative regulatory protein is not able to bind, the regulated genes are constitutively expressed. (A defect in a positive regulation site would lead to reduced expression of the regulated genes because the positive regulator could not bind to the DNA to act.)

b) Strain i) would have inducible *emu1* and constitutive *emu2* expression, while strain ii) would have inducible *emu1* and *emu2* expression

c) Strain i) would have inducible *emu1* and *emu2* expression while strain ii would have inducible *emu1* and constitutive *emu2* expression.

15-9. b) When a positive regulator is inactivated, there is no expression from the operon.

15-10. a) 1 b) 6 c) 2 d) 4 e) 5 f) 3

Look first for striking patterns that might indicate what the mutation is. For example, mutant 6 that shows expression of *lacZ* when combined with any of the other mutations. Of the possible choices, this could only be an o^c mutation. Strains containing mutation 5 and any of the mutations other than mutation 6 also has a very consistent pattern. There is never expression of *lacZ* except when combined with mutation 6 (the o^c mutation). Mutation 5 therefore shuts down the other copy of the operon in addition to its own copy, which can be explained by a superrepressor mutation. Looking at the remaining types of mutations, the inversion of the *lac* operon (d) should not have an effect on expression and should not influence expression from other copies of the operon. Mutation 4 leads to inducible *lac* expression except when combined with mutation 5, the super repressor. The inversion that does not include *lac I, p,* and *o* should not show expression because the regulatory region is now in the opposite orientation from the genes. It also should not influence expression from the other mutant copy. Looking at the patterns for mutations 2 and 3, mutation 3 does not show expression except when combined with 4, so fits with inversion. Mutation 2 is a *lacZ* mutation.

15-11. a) Conditions 1 and 2 (all wild-type genes in the presence and absence of arabinose) are evidence that arabinose induces expression of the *araBAD* genes.
b) Strains and conditions 2 and 4 are sufficient to conclude that *araC* encodes a positive regulator because the mutant version of *araC* results in no synthesis of the three gene products in the presence of arabinose.

15-12. a) repressor (LacI protein)
b) no proteins bound
c) CAP-cAMP complex

15-13. a) 4 b) 6 c) 7 d) 2 e) 3 f) 5 g)1

This problem can be approached either by starting with the expression data and assessing what types of mutations could produce that pattern or starting with each mutant type and matching it with an expression pattern. Starting with the latter approach, the superrepressor mutant (a) would show no expression under any conditions and therefore could be either mutant 3 or 4. The operator deletion (b) would result in constitutive high level expression under glycerol or lactose growth conditions but low expression under lactose + glucose (because it would be catabolite repressed) and could therefore be mutant 5 or 6. The amber suppressor tRNA would have no effect on its own and therefore is mutant 7. The defective CAP-cAMP binding site would produce the same low levels of expression under lactose or lactose + glucose since the CAP-cAMP complex cannot bind to increase expression. Mutant 1 or 2 could contain a mutation in the CAP-cAMP binding site. The nonsense mutation in β-galactosidase would result in no expression (mutant 3 or 4). It would be suppressed by mutation 7 (amber suppressor tRNA) in the same cell so mutant 3 contains the nonsense mutation in β-galactosidase (*lacZ*) gene. A nonsense mutation in the repressor gene would be constitutive when grown in glycerol or lactose, so must found in mutant 5. The defective *crp* gene (g) would have the same expression profile as the defective CAP-cAMP binding site so could be either 1 or 2. These two are distinguished by the fact that the CAP-cAMP binding site mutation would be present in the F'lac and would therefore have the same profile as the mutation present on the chromosome, while the *crp* mutation would not be picked up on the F'lac (because it is not part of the *lac* operon) and would not have the same profile. Mutant 1 is the *crp* mutation.

15-14. a) There is no expression because the promoter is deleted and transcription cannot occur.

b) There will be partially constitutive *trpC* and *trpE* expression in the presence of tryptophan. The repressor is defective, so transcription initiation is constitutive but the attenuator is normal so will result in less expression under tryptophan growth. When tryptophan is not in the media, this operon will be fully induced.

c) The operon is repressed (in the absence or presence of tryptophan) because tryptophan cannot bind to the repressor to release it from the operator.

d) Expression of *trpC* and *trpE* is completely constitutive because both the repressor and attenuator are mutant.

e) There is inducible expression of *trpC*, partially constitutive expression of *trpE* because the operator connected to the *trpE*$^+$ is defective (but the attenuator is still functional).

f) There is inducible expression of *trpC* but no expression of *trpE* because the promoter in cis is defective

g) Expression of *trpE* is fully constitutive because the att site is in cis to the *trpE*$^+$ copy, but *trpC* expression is partially constitutive because the *trpC*$^+$ gene is next to a functional att site.

15-15. If the three genes make up an operon, they are cotranscribed as one mRNA and only one band should be seen after hybridizing a probe from any of the three separate genes with mRNA from the same. The band will be the same size using a probe from any of the genes and will be large enough to include the RNA copy of all three genes together. If the genes are not part of an operon, each would be transcribed separately so there would be three hybridizing bands.

15-16. a) A, C, and D are turned off when compound Z is added. You could think of this operon (three proteins from three coregulated genes) as encoding genes required for biosynthesis of compound Z.

b)   A- structural gene

B-repressor

C- structural gene

D- structural gene

E- attenuator

F- operator

G- promoter

A, C, and D are structural genes needed for the biosynthesis of compound Z. Compound B is constitutively produced so is not part of the transcription unit including the structural genes. When B is mutant, there is only a partial turn-off of genes. B could be a repressor that shuts off synthesis of A,C,D genes when compound Z is present, but the shutting down of transcription has a second component - an attenuator. This explains why there is still some turn down of transcription even when B is mutant. The nonsense mutation in C affects not only the presence of protein C, but also A and D, so C must be transcribed before these two genes. Nonsense mutation in A causes no A or D to be produced while nonsense mutation in D only affects the synthesis of D. The order of genes is C A D. Deletion of site G results in no expression of any of the three structural genes, so is probably a promoter. Looking at the remaining deletion mutations, removal of either E or F lead to the same reduced repression. This could occur if there were a repressor binding to an operator and an attenuator site in this operon, each being responsible for a 10-fold repression when compound Z is available. Sequence E codes for a small peptide and since small peptides are part of attenuator sites in amino acid biosynthetic operons, we can assume that deletion of E and adjacent DNA would delete an attenuator site. Site F then is the operator site. The operator and promoter are first, then the attenuation region including the leader peptide, then the structural genes. B is not cotranscribed with the other genes, so we do not know its location and it is shown on the map below in parentheses.

	operator/promoter		attenuator		structural genes		
			peptide	leader			
(B)	F	G	E		C	A	D

15-17. Ribosomes would pause at the His codons (CAC or CAU) in the sequence because the tRNA ^{His} would not be completely charged with histidine.

15-18. Any mutations that affect presence of a functional LamB protein would be resistant to lambda. This would include:

lamB point mutations or deletions

malT mutations that prevent positive regulation

mutations in site to which *malT* binds

mutations in the *crp* gene (encoding CAP protein)

mutations in CAP-cAMP binding site of the *malK lamB* operon

malK lamB promoter mutations

15-19. a) Operator begins on the left site after the endpoint of deletion 1 and before the endpoint of deletion 5. The right endpoint cannot be determined by this data.

b) The deletion may have removed bases within the promoter that are necessary for transcription to begin.

15-20.

15-21. Mutations in O_2 or O_3 alone have only small effects in synthesis levels and would therefore be difficult to detect in the screens.

15-22. a) If the screening was done for Lac$^+$ only, some mutations might be compensating mutation in the CAP-cAMP binding site for that operon specifically. Demanding both Mal$^+$ and Lac$^+$, suppressor mutations would be more general, affecting all CAP-cAMP binding or activity.

b) The α subunit of RNA polymerase interacts directly with CAP protein.

15-23. The fusion joint has put the protein coding region of your gene in the same reading frame as the *lacZ* gene.

15-24. positive regulator. If a loss of function in a regulator leads to expression, the regulator protein was needed in a positive way to express the genes.

Chapter 16   Gene Regulation in Eukaryotes

Synopsis

This chapter describes different ways in which gene expression is regulated in eukaryotes. Regulation of transcription is a major means of controlling expression. Regulatory proteins (activators and repressors) bind to sites in the DNA near the gene to turn up or down expression. The low level of basal transcription of a gene requires a set of transcription factors binding at the promoter. Gene expression that occurs in only certain cells or is specific for a stage in development or occurs in response to external factors is mediated by additional proteins binding near the coding region. Characteristic motifs that are important for DNA binding or interaction with RNA polymerase or with other activator proteins can be recognized in regulatory proteins that bind to DNA.

Other controls of gene expression include alternate splicing events (producing different mRNAs from the same gene), mRNA stability, control of translation, different packaging of region of DNA in chromatin, and protein modifications.

Be prepared to:

- distinguish between positive and negative regulators based on the effects of mutations
- design experiments to determine the sequence(s) of a regulatory region of a gene that are important for basal transcription and which sequences are important for tissue specific expression
- interpret expression of a gene using a reporter gene fused to a regulatory region
- identify interacting proteins in regulation of a gene
- interpret expression of a paternally or maternally imprinted gene

Problem-solving tips

- When a regulatory region of gene "X" is cloned next to the *lacZ* gene and reintroduced into eukaryotic cells, the expression of β-galactosidase will reflect expression patterns of gene "X".
- Deletions of sites to which activators bind will lead to lower expression; deletions of sites to which repressors bind lead to increased expression.
- Several regulatory proteins can interact to regulate one gene.
- Mutations in regions of proteins that interact with another protein or a site will only affect that one specific function.
- DNase hypersensitive sites indicate that the DNA is in a less compacted form and is more available for transcription.

- For maternally imprinted genes, the copy of the gene inherited from the mother will not be expressed in either her sons or daughters. For paternally imprinted genes, the copy of the gene inherited from the father will not be expressed in either his sons or daughters.

Solutions to Problems

16-1. a) 8 b) 5 c) 1 d) 7 e) 3 f) 2 g) 6 h) 4

16-2. a) eukarotyes b) prokaryotes and eukaryotes c) eukaryotes d) prokaryotes e) prokaryotes and eukaryotes

16-3. Introns are spliced out of a primary transcript and a ribonuclease cleaves the primary transcript near the 3' end to form a new 3' end to which a polyA tail is added.

16-4. a) iii b) i c)ii

16-5. a) pol III b) pol II c) pol I

16-6. With a promoter sequence only cloned next to *lacZ*, there should be a low level of β-galactosidase expression. Promoter + enhancer will give high levels of expression while enhancer alone would show no expression. The promoter must be located in the last M-H fragment since each of the clones showing low (basal) levels of transcription contain this fragment and any clones not containing this fragment do not produce β-galactosidase. The enhancer is present in the first H-M fragment since that is found in the two clones (sixth and eighth) showing high expression. Clone 2 that contains this fragment does not show high levels of expression because the promoter is not included in the clone.

16-7. a) DNA binding b) DNA binding c) dimer formation d) transcription activation e) DNA binding

16-8. a) base pair -83 to -65.

The fact that the deletion to -85 has full activity indicates that the binding site is after -85. The binding region must begin after the endpoint of the first deletion (-85) and within that region there is a short sequence that is almost repeated (between −83 and −65), so you might predict that would be the monomer binding site (which must be repeated to get active dimer binding) and therefore should be included.

b) The likely consensus for the monomer binding would be: CTG(C or T)G

16-9. You could do a Northern hybridization analysis using probes from each of the three genes versus mRNAs separated by size using electrophoresis and transferred to a filter for hybridization. The expected result is different signals corresponding to three different transcripts. You would not see a band hybridizing that would be large enough to contain the coding information for all three genes.

16-10. A *GAL80* mutation in which the protein is not made or is made but cannot bind to Gal4 would prevent repression and lead to constitutive synthesis. A *GAL4* mutation in the portion of the protein to which Gal80 normally binds to repress Gal4 action should be constitutive. (A site mutation occurring in the sequence to which Gal4 binds should also lead to constitutive synthesis.)

16-11. If Id acts by quenching, it interacts with MyoD whereas if it blocks access to an enhancer (site in the DNA), it binds to DNA. Experimentally, you could look for binding to the regulatory DNA of a gene regulated by MyoD.

16-12. You could prepare RNA from sporulating cells and make cDNA copies of the all RNAs. The cDNAs then could be cloned into a vector to generate the library representing genes expressed during sporulation.

16-13. RNA polymerase transcribes naked eukaryotic DNA at high levels in vitro. In contrast, transcription of genes packaged in chromatin occurs at a low basal level and requires transcription factors to be transcribed at higher levels.

16-14. The gene is more likely to be transcribed in liver cells, based on the DNaseI profile, than in muscle cells. Liver cell DNA has a DNaseI hypersensitive site that results in the two bands, 16 and 4 kb in size. The site must not be exposed in the muscle cells, and this is indicative of lower transcription activity.

16-15. b.

16-16. a)2 b)4 c)1 d)3

16-17. a) not expressed. The gene is paternally imprinted and the imprint is set during gametogenesis, so male I-1 will not express the gene in germ cells.

b) not expressed. The gene inherited from her father was turned off in his germ cells, so the somatic cells arising after fertilization will contain an inactive copy from the father.

c) expressed. The imprint from the father is erased during gametogenesis and a female will not inactivate this gene (because it is a paternally, not maternally imprinted gene).

d) not expressed. The gene inherited from his father was turned off, so the son's somatic cells will not express the gene.

e) not expressed. The imprint from the father is erased during gametogenesis but the male will re-imprint the gene (inactivate expression).

f) expressed. The son of the daughter inherited the non-imprinted version of the gene from his mother, so the gene is expressed in somatic cells

g) not expressed. In the germ cells, the male imprint is re-established.

16-18. The mRNA for this gene is made in all tissues tested, while the protein is made in only muscle cells. Expression of this tissue specific gene is regulated at the level of translation.

16-19. a) 9:7 ratio of blue to white flowered plants.
A=transcription factor 1; B=transcription factor 2. If both transcription factors are required to get expression, only flowers with the *A-B-* genotype will produce blue flowers.

AaBb x AaBb	9/16*A-B-*	blue
	3/16*A-bb*	white
	3/16*aaB-*	white
	1/16*aabb*	white

b) 15:1 ratio of blue to white flowers. In this case, three of the four genotypic classes from a dihybrid cross would produce blue flowers.

9/16 *A-B-*	blue
3/16 *A-bb*	blue
3/16 *aaB-*	blue
1/16 *aabb*	white

16-20. The different cDNAs could be due to a different start site for transcription in the two cell types or due to alternate splicing in the two cell types.

16-21. The 5' and 3' regions could be cloned at the 5' or 3' ends of a reporter gene (for example, *lacZ*) that is transformed back into *Drosophila* early embryos to see if either sequences cause translation of the reporter protein at the appropriate times during development.

16-22. The protein must need modification (for example, phosphorylation) to become active.

Chapter 17   Cell-cycle Regulation and the Genetics of Cancer

Synopsis

This chapter describes characteristics of the eukaryotic cell cycle and the cancer phenotype. Genetic analysis of the regulation of the cell cycle has lead to our understanding of the genetic basis of cancer. Cancer is an uncontrolled growth of cells – the cell cycle is no longer correctly regulated. There are genes whose products are necessary for the cells to proceed in the cell cycle and genes whose products monitor progress of the cell cycle, stopping the cycle if there is something amiss. Signals for beginning cell division are relayed through a series of steps (signal transduction). A series of mutations, each one either inherited or newly arising in a somatic cell lead to a cell dividing out of control. Inherited predisposition to particular cancers is due to a mutation in one of the genes controlling cell cycle growth.

Be prepared to:

After reading the chapter and thinking about the concepts, you should be able to:

- distinguish between oncogenes and tumor suppressor genes based on the behavior of cells containing a mutation in a gene
- describe the role of growth factors and the pathway of signaling that leads to the start of the cell cycle
- describe the role of cyclins in regulating the cell cycle
- explain inherited predisposition to a non-scientist

Problem-solving tips:

- Think about gene regulation, cell signaling and how they all might fit together in a system for regulating the cell cycle.
- Remember that interacting proteins will have regions of genes in which mutations affect the ability to the proteins to interacting
- Use your knowledge of molecular techniques (hybridizations, PCR, transcription analysis, etc) to design experiments
- Multiple occurrences of cancer and/or early onset of cancer are suggestive indicators of inherited predisposition alleles, but further experiments are necessary to identify the mutant alleles especially if that type of cancer is very prevalent in the population.

Solutions to Problems

17-1. a) 7 b) 6 c) 8 d) 2 e) 1 f) 9 g) 3 h) 5   i) 4

17-2. a) M (mitosis)

b) M (mitosis)

c) S phase

d) G_1 phase

17-3. Grow mutagenized yeast cells at 30°C and replica plate colonies onto two sets of plates- one incubated at 30°C and the other at 23°C. Genes necessary for the cell cycle should be essential, so cold-sensitive mutants defective in cell cycle genes should die at the non-permissive temperature (23°C). Those mutants that are cold-sensitive can be studied more extensively by growing a liquid culture of the strain at permissive temperature, shifting to the non-permissive temperature and watching the cells microscopically to see if they stop with the same phenotype. If they have the same phenotype when they stop dividing, this indicates the stage in the cycle at which they stop.

17-4. Three complementation groups and therefore three genes. Mutant 1 does not complement mutants 4,5 so these three are in one complementation group. Mutant 2 does not complement mutant 8, so these make up a second complementation group. Mutant 3 does not complement mutants 6,7, or 9 so these are in a third complementation group.

17-5. Chromosomes are visible during mitosis (although condensation begins during prophase) and mitosis is 1/24 of the cycle, so 1/24 or 4% of the population of cells are in M.

b) The remaining proportion of cells, 96% (23/24) are in interphase (G1, S and G2 combined).

17-6. There can be cyclical transcription, translation, or posttranslational modifications that lead to active proteins.

17-7. a) False (CDKs need cyclins for activity.)

b) True

c) False (Checkpoint monitoring proteins check for aberrant cell cycle events.)

17-8. a.2 b.3 c.1

17.9 a) Examples of external molecules would be any of a number of growth factors.

b) A molecule inside the cell could be a cyclin, cyclin dependent protein kinase, any molecule in the signal transduction pathway, receptor or molecule that transmits a second signal.

17-10. a) A mutant RAS that stays in the GTP-bound state is activated and will cause the cell to continue dividing.

b) If the mutant stayed in the GDP-bound form, it would not be able to respond to signals that use the RAS pathway.

17-11. c. e. a. b. d.

17-12. a) T antigen binds to p53 and prevents it from acting in cell cycle regulation. By supplying excess p53 from the high level promoter, there is enough now in the cell, so the effect of the T antigen is minimized.

b) Mutants 1 and 2 can no longer rescue the cells from the T antigen effect so must be mutations in the part of p53 that functions in cell cycle control.

c) Mutation 3 still rescues the cells from T antigen so it must be a change in the DNA sequence in a region of p53 necessary for cell cycle control. We know from this result that this base is not necessary for T antigen interaction.

17-13. Four characteristics of cancer are uncontrolled growth, genomic instability, potential for immortality, ability to invade and disrupt local and distant tissues.

17-14. a) Amplification of a specific sequence could be observed using hybridization to DNA from tumor and normal cells. You would want to use two different probes- one representing the specific sequence that you are analyzing and the other representing another sequence to use as a control (assuming the latter sequence is not amplified too).

b) Gross rearrangements could be seen by karyotype analysis. You would look for alterations in the banding patterns.

17-15. We know that some of the environmental agents that are implicated in increased cancer risk caused increased level of mutations, so this fact is consistent with the fact that mutations in genes are necessary to cause cancer. (The inherited mutations that lead to predisposition are just one of several mutations that must occur within a cell to lead to cancer.) The environmental and inherited effects are therefore very related.

17-16. To assess the role of genetic differences, you would need to keep other factors (for example, diet) as constant as possible. You could look at the incidence in Indians and Americans who have similar diets. This is often done by studying cancer rates in immigrants. For example, people who moved from India to Toronto could be compared with Toronto natives.

To assess the role of diet, studies can be set up within the immigrant population (or the Toronto native population) where an experimental group gets a diet resembling the other ethnic group and a control group continues with their normal diet.

17-17. v, ii, i or iv or iii, vi

17-18. ii, iv, i, iii

17-19. c) is unlikely to be associated with cancer because a loss of function of one copy is not a dominant mutation that is needed to convert a proto-oncogene into an oncogene.

17-20. a) The appearance of so many affected individuals in one family suggests an inherited mutation is contributing to development of the disease. Early age of onset is very often a clue that a germline mutation may have been inherited, but it is not an absolute predictor.
b) The inheritance pattern suggests autosomal dominant inheritance of a predisposing mutation but individuals II-2, and either I-1 or I-2 we would predict carry the mutation. Since these individuals are not among the high coffee consumers, the combination of consumption of the special coffee and a particular genotype may be an environmental factor that results in the phenotype of colon cancer. A substantial number of individuals who drink coffee are not affected (for example III-8, III-9, III-10, III-11, III-12) , so coffee alone does not appear to have a substantial effect.

17-21. If there is a dominant predisposing allele segregating, individual II-2 carries the mutation since male III-2 has the predisposition. Because two individuals in the second generation carry the mutation, one of the parents from generation I must have passed on the allele, so either I-1 or I-2 also carries the mutation. (If only one member of generation 2 carried the allele, it would be possible that the mutation arose in development of a germ cell in a generation 1 individual. Then the somatic tissues of that parent would not contain the mutation.)

17-22. (d) Hybridization with allele-specific oligonucleotides would be the most rapid method for identifying a specific point mutation. If PCR amplification (a) were done, it would be necessary to then sequence the DNA that was amplified to determine if the mutation were present. If the mutation had been one in which a restriction site was altered, restriction enzyme digest followed by Southern blot could be used.

Chapter 18   *Saccharomyces cerevisiae:* A Genetic Portrait of Yeast

Synopsis:

 Saccharomyces cerevisiae is a model eukaryotic cell for studying basic cell processes. The following features of yeast genetics and molecular biology make it a powerful model system.

- Easy to screen for rare mutations.

- Tetrad analysis allows direct analysis of the products of meiosis in one cell. $2^+:2^-$ segregation indicates that a mutation in a single gene is causing the phenotype.

- Mutants can be isolated in haploids or diploids. Recessive mutations are immediately recognized in haploids. If mutations are being sought in an essential process, mutagenesis can be done in diploids. The recessive mutation is identified in the haploid cells after sporulation. Sporulation (meiosis) can be induced in a culture by specific growth conditions.

- Yeast cells are easy to transform.

- Several plasmid vectors have been developed that allow easy growth in either yeast or bacteria. Artificial linear chromosomes allow the study of chromosome structure.

- Yeast chromosomes can be separated by size using pulse-field gel electrophoresis.

- Gene replacement allows a researcher to change a gene specifically by transforming with a replacement copy (engineered variant) that is forced to recombine and replace the original copy.

Be prepared to:

After reading the chapter and thinking about the concepts, you should be able to:

- predict the effects of mutations in the components of the mating system (production of pheromone, recognition of pheromone, signal transduction) and be able to apply these analytic principles to other processes.

- test mutations to determine if they are in the same or different genes (complementation analysis); determine if mutations in different genes act in the same or different pathways (epistatic analysis)

- think about how genetic analysis is done in this organism including: isolation of mutants, characterization of mutations (complementation and epistasis analyses), mapping a gene, cloning a gene, making a mutant cloned copy to reintroduce into the genome

Problem-solving tips:

- Reason through the problems; get used to making hypotheses and interpreting how results fit a hypothesis

- Some of the longer problems in this chapter are real research problems that were carried out essentially as described to you. They require that you integrate your knowledge of classical genetics, molecular genetics, genome analysis. This is a good time to try out thinking like a geneticist. Think about how you would discover the genes involved and how their gene products may act- as regulators, messengers, structural components, etc.

Solutions to Problems

18-1. a) 3 b) 4 c) 5 d) 7 e) 2 f) 1 g) 6

18-2. Assuming an average protein coding region size of 500 amino acids, there would be 1500 nucleotides per gene. 6000 open reading frames (protein coding regions) $\times$ 1500 = 9,000,000 bp or 9,000 kb in protein coding DNA. This value is 3/4 of the total genome size (9000/12,000).

18-3. You could isolate DNA from cultures of each of the ten clones, use pulse-field gel electrophoresis to separate the chromosomes by size and compare the sizes of chromosome XII in each isolate.

18-4. a) If the histidine biosynthesis genes are coregulated, the mRNA for each gene should be induced by the same conditions. For amino acid biosynthesis genes the genes are usually turned on by starvation for the appropriate amino acid. Cells would be grown with and without histidine in the media, RNA isolated and separated by electrophoresis, transferred to filter where they could be probed with DNA representing each of the biosynthetic genes. If they were coregulated, the RNA for each would probably be induced by starvation for the amino acid. (Note: Coregulation is not the same as cotranscription as seen in prokaryotic operons. You would not expect the genes to be cotranscribed as one mRNA, because that is found almost exclusively in prokaryotes.)
b) You might expect to find similar DNA sequences in the regulatory regions to which transcriptional regulators would bind to regulate expression.

18-5. a) You could separate all the chromosomes by pulse-field gel electrophoresis, transfer to filter and do a hybridization using probe DNA containing a subtelomeric repeat sequence.

b) You could use the cloned piece of DNA as a probe in a hybridization with a filter containing DNA from yeast chromosomes that had been separated by pulse field gel electrophoresis.

18-6. a) The $2^+:2^-$ segregation pattern is characteristic of a monohybrid segregation of the alleles of a gene from a heterozygote during meiosis. Because the analysis is done in *Saccharomyces cerevisiae*, a fungal species in which the four meiotic products are packaged together, the ratio is easily seen. (The 2:2 segregation also indicates that the strain is heterozygous).

b) The screen identified the chromosome segregation defect in a diploid cell (which we know from the sporulation results is heterozygous). The phenotype seen in a heterozygous diploid is the dominant phenotype, so in this case the mutant allele is dominant.

18-7. The chromosomes containing the MAT locus could align in any combination during meiosis, leading to the following combinations of a and α alleles in the diploid spores: *a/a* which has an *a* mating phenotype, *α/α* which has an α mating phenotype and *a/α* which is a nonmater (sterile).

18-8. a) ii, iv, v

b) i, ii, iv, vi

c) i, iii, iv, vi

18-9. a) In an α cell, α1 + Mcm1 protein act as a positive regulator of α-specific gene transcription; α2 + Mcm1 protein act to repress *a*-specific genes. In an *a* cell, Mcm1 protein alone transcribes *a*-specific genes. In an a/α strain, Mcm1 has no effect.

b) In the α *Mcm1*⁻ strain, there would be no transcription of α-specific genes and no repression of *a*-specific genes, so the strain behaves and mates as an *a* strain. In the *a Mcm1*⁻ strain, there would be no *a*-specific gene expression (for example, no *a* pheromone), so the strain would be sterile. The *Mcm1*⁻ mutation would have no effect in an a/α strain. It would be sterile as is the *Mcm1*⁺ a/α strain.

18-10. a) complementation

b) The 100 haploid spores that grew on minimal medium must contain two wild-type alleles. The $1^+/1^-$ $2^+/2^-$ diploid, for example, would produce spores with four different genotypes: 1^+2^+ (which grows on minimal), 1^-2^+, 1^+2^-, 1^-2^- (none of which grow on minimal media) if the genes are unlinked. One-quarter of each type of spore would be produced, so there would be 100 1^+2^+ spores out of 400 spores. Mutations that are in the same gene or are in closely linked genes will only produce the 1^+2^+ and 1^-2^- by recombination between the mutations- a low frequency event dependent on distance between them. The combinations of mutations that only produce a low number of spores (< 5) must be closely linked mutations.

c) There are 4 complementation groups: 7,3; 4,5; 2,6; 1. These are identified by looking at the growth of the diploids on different intermediates.

d) sleepan $\longrightarrow$ happan $\longrightarrow$ dopan $\longrightarrow$ sneezan

$\qquad$ 7,3; 4,5 $\qquad\qquad$ 1 $\qquad\qquad$ 2,6

Because none of the mutants grew on sleepan, this is the earliest intermediate. On happan, 7,3; 4,5 grow but because these alleles are in two different complementation groups they represent two different genes. The enzyme that converts sleepan to happan is therefore composed of at two different subunits (heterodimer). Dopan is the next intermediate with the gene identified by mutation 1 producing dopan. The gene identified by mutations 2 and 6 encodes the enzyme to convert dopan into sneezan.

18-11. a) You could do a Southern hybridization using the *SUC2* DNA as a probe in a hybridization with a filter containing genomic DNA from other *Saccharomyces* strains.

b) pseudogenes

c) There are two transcripts made from the *SUC2* locus. The smaller one is produced when glucose is present or absent; the larger one is present only when glucose is absent. The larger transcript requires the *SNF1* gene product but the smaller one does not require *SNF1*. The larger transcript seems that it encodes the functional transcript.

18-12. a) A complementation test would be done by crossing strains to each other and examining the phenotype in the diploid. (The mating type of the haploids crossed to each other would have to be opposites so that *a* and α cells were mated.)

b) Three loci. Mutations 1, 2, 4 are in the same gene; 3, 5 are in another gene; 6 is in another gene.

c) Mutations 1 and 3 are in genes in a different pathway; 3, 6 are in different pathways; and 1 and 6 are in the same pathway.

d) Isolate RNA from cells under normal growth conditions and from cells exposed to UV. Separate the RNAs by gel electrophoresis, transfer to filter and hybridize using specific gene probes to determine if expression of the genes is regulated by the UV exposure.

e) Compare the expression of *lacZ* in your library of clones before and after exposure to DNA damage agents. Any of the cells that show increased expression of β-galactosidase after exposure would contain clones in which a regulatory region of a UV-inducible yeast gene had been cloned next to the *lacZ* gene.

Chapter 19   *Arabidopsis thaliana:* Genetic Portrait of a Model Plant

Synopsis

     Arabidopsis thaliana serves as a useful plant model for genetic and molecular analyses. While many of the basic cellular mechanisms are the same in plants and animal cells, the events during development in plants are quite different from animals. Even the unique developmental events in plants often have underlying mechanisms that are analogous to those found in animals, fungi, etc. The following list of characteristics of *Arabidopsis* indicates why it is a useful model for plant development and point out genetic characteristics and manipulations that are unique to this plant.

- Fast generation time compared to other plants.
- Large number of seeds per plant (good for mutant screening).
- Self-fertilization occurs or crosses can be manipulated by the researcher.
- Mutant screens can be done using T-DNA or other transposable elements to create mutations.
- Very little repetitive DNA is present in the genome (This contrasts with other plants where the genome size is very large due to repetitive DNA).

Be prepared to:

After reading the chapter and thinking about the concepts, you should be able to:

- interpret experimental data using fusions to study expression, molecular markers, mutagenesis, characterization of mutants, PCR, hybridizations, genome data
- describe steps in a search for T-DNA tagged mutant
- think about how genetic analysis is done in this organism including: isolation of mutants, characterization of mutations (complementation and epistasis analyses), mapping a gene, cloning a gene, making a mutant cloned copy to reintroduce into the genome

Problem-solving tips:

- Remember that a seed contains a plant embryo. When seeds are mutagenized, you are not mutagenizing gametes. Gametes are produced in the reproductive structures of an adult plant.
- Some of the longer problems in this chapter are real research problems that were carried out essentially as described to you. They require that you integrate your knowledge of classical genetics, molecular genetics, genome analysis. This is a good time to try out thinking like a

geneticist. Think about how you would discover the genes involved and how their gene products may act- as regulators, messengers, structural components, etc.

Solutions to Problems

19-1. a) 3 b) 1 c) 4 d) 5 e) 2

19-2. a) The shorter generation time of *Arabidopsis* allows the geneticist to study many more generations in less time.
b) The very large number of seeds produced is an advantage for a geneticist searching for rare mutations.
c) The small genome size is advantageous for molecular analysis, in particular for cloning genes. There is very little repetitive DNA to have to deal with.

19-3. Most of the extra DNA in tobacco and pea is repetitive DNA which presumably is not necessary for the physiology of the plant. Repetitive DNA does not encode genes so does not contribute to the proteins available in the cell.

19-4. a) Plants of different ecotypes that show the extremes (large compared to small seed set) could be crossed to produce the hybrid F_1 that is then crossed to produce the F_2. If the F_2 shows discrete phenotypes, there is a single gene that determines the trait. If the F_2 plants have a range of phenotypes in a continuum, the trait is determined by several genes.
b) Different ecotypes contain different forms of molecular markers that can be used to identify regions containing genes involved in determination of the phenotype (quantitative trait analysis).

19-5. a) Endosperm is 3N and is formed by fertilization of one sperm nucleus and two 1N nuclei in the ovule; the genotype will be like that of the embryo- a combination of genetic information from both parents (from an outcross).
b) The zygote is 2N and is formed by fertilization of one sperm nucleus with the 1N egg; the genotype will be a combination of genetic information from both parents.
c) The embryo sac cells are 2N and they are part of the parent plant; have the genotype of the parent plant, not the embryo.

19-6. A geneticist usually wants to study mutations that are in the germline and can therefore be maintained as stocks for further study. The first generation mutations are in the somatic tissue. The cells of the embryo that are in the seed and will divide and develop into the plant are the cells that were mutagenized.

19-7. a) 2 b) 3 c) 1

19-8. a) iv, iii,i, ii
b) The DNA you would transform into the cell would have your gene of interest with a selectable gene (e.g., NPTII) inserted into the coding sequence.

19-9. You could do manipulated crosses where you collect pollen from the sterile plant and use it to fertilize a wild-type plant. The reciprocal cross would also be done: pollen from a wild-type plant crossed with the sterile mutant. From the results you would know if the defect was in the male or female reproductive structures.

19-10. To look for an homolog for a gene in other plants, make DNA from the petunias, snapdragons, potatoes; cut DNA with restriction enzymes, separate by electrophoresis, transfer to filter paper to do a Southern hybridization using the *APETALA* gene from *Arabidopsis* as a probe.

19-11. a) You could examine the gravitropic response in an auxin mutant (defective in auxin production). If the gravitropic response is lacking in an auxin mutant, the response must be auxin mediated.
b) There are a few different ways you could examine the distribution. You could grind up tissue from different parts of the plant, make DNA and hybridize using an auxin-regulated gene as a probe. For a more accurate picture, you could use DNA for an auxin regulated gene as a probe against tissue from the plant (an in situ hybridization) Or, you could make a fusion between an auxin regulated gene and a reporter gene such as GUS or luciferase (emits light when given substrate luciferin) and look at the distribution of expression of the reporter gene in the intact plant.

19-12. a) The *Ctr1⁻* constitutive mutation in a signaling pathway should not be affected by what is happening in the ethylene biosynthesis, so there would still be constitutive expression of the ethylene response.

b) Because inhibitors of biosynthesis affect these mutants, the mutations are in a gene(s) in the biosynthetic pathway.

c) You could look for additional copies of the *ETR1* gene by Southern hybridization using the *ETR1* gene as a probe in a hybridization to a blot containing genomic *Arabidopsis* DNA.

d) Both *EREBP* and *APETALA* are DNA binding proteins so might share some similar features (types of motifs) in the protein.

e) To control expression of the ethylene response genes, you would need to make a fusion of an inducible regulatory region with the ethylene response gene you want to induce. The inducible regulatory region would have to be one that you could turn on in the plant cells where you wanted expression.

Chapter 20   *Caenorhabditis elegans:* Genetic Portrait of a Simple Multicellular Organism

Synopsis:

The nematode *C. elegans* has become a very important model for genetic analysis in multicellular organisms based on the following characteristics:

- Ability to grow large numbers of worms quickly and screen for mutants.
- Self-fertilization occurs in hermaphrodites and therefore makes the identification of recessive mutations easier and more controlled than in animals where matings would have to be set up between strains containing the same mutation in different genetic backgrounds.
- Small genome size means that cloning and genome analysis is easier.
- Basic animal functions are present in the worm.
- Ability to view internal organs and cells in a living worm via microscopy.
- Known, precise lineage of cells during development.
- Ability to kill specific cells via laser ablation.
- Free duplications (extrachromosomal pieces of DNA) can be introduced to create mosaic worms in which some cells contain the free duplication.

Be prepared to:

After reading the chapter and thinking about the concepts, you should be able to:

- set up crosses to construct strains of worms with specific genotypes
- interpret results of selfing in a hermaphrodite and outcrosses between hermaphrodite and male.
- think about how genetic analysis is done in this organism including: isolation of mutants, characterization of mutations (complementation and epistasis analyses), mapping a gene, cloning a gene, making a mutant cloned copy to reintroduce into the genome

Problem-solving tips:

- When a hermaphrodite population is mated with a male population, the progeny will arise from self-fertilization of hermaphrodites as well as the outcrosss of a male to a hermaphrodite.
- Males (XO) arise by non-disjunction. Hermaphrodites are XX.
- Some of the longer problems in this chapter are real research problems that were carried out essentially as described to you. They require that you integrate your knowledge of classical

genetics, molecular genetics, genome analysis. This is a good time to try out thinking like a geneticist. Think about how you would discover the genes involved and how their gene products may act- as regulators, messengers, structural components, etc.

Solutions to Problems

20-1. a) 2 b) 5 c) 4 d) 7 e) 6 f) 1 g) 3

20-2. Five times the genome size (100Mb) is 500 Mb or 500,000 kb. With an average cosmid size of 50 kb, you would need 10,000 clones to make sure that you have a coverage of 5 times the genome.

20-3. Hermaphrodites have the advantage of self-fertilization. It is therefore easier to recover recessive alleles than if there was only uncontrolled outcrossing. (In other animals, you would require an additional generation of crosses to obtain zygotes homozygous for a recessive mutation because it is necessary to first get the mutation in individuals of both sexes.) In addition, mutants with some pretty severe defects (such as paralysis) can be isolated because the worms do not have to move to mate.

20-4. a) iv, ii b) i, ii c) iii

20-5. b) would be expected if trans-splicing occurred.

20-6. Because the induction of cell fate by an anchor cell occurs during development, you would ablate the anchor cell during development, at a time just prior to when the induction is believed to occur - in other words, preceding those cell divisions in development after which the fate of the potential anchor cell influenced cells has been established. As described in the text, this would be during the L3 larval stage.

20-7. Increased *lin-4* mRNA would probably reduce the Lin14 level further and this could lead to cells differentiating at an inappropriate time since Lin14 protein concentration acts as a clock during development. The cells would differentiate inappropriately earlier than in wildtype. This should be similar to a *lin-14* loss-of-function mutation.

20-8. You could probe DNA from the mutant and wild-type with an the *unc-22* cloned DNA. If there was an insertion of a Tc1 element, the fragments that hybridized in the mutant and wild-type would be different. You would then want to confirm that there is a fragment that contains both Tc1 and *unc-22* DNA to show that the insert is within the *unc-22* gene. Alternatively you could use PCR with one primer consisting of a Tc1 sequence and the other being an *unc-22* sequence. If a fragment is produced using these primers, the Tc1 element must have transposed in the vicinity of the *unc-22* gene.

20-9. a) Cross the wild-type males to the homozygous *dpy-10⁻/dpy-10⁻* (or homozygous *daf-1⁻/daf-1⁻*, it doesn't matter).

b) From the first cross, the wild-type male progeny would be crossed to the *daf-1⁻/daf-1⁻* (or *dpy-10⁻/dpy-10⁻*) hermaphrodites.

c) The final cross is the self fertilization of the wild-type hermaphrodites from cross 2. The wild-type hermaphrodites from cross 2 are distinguished from the *daf-1⁻/daf-1⁻* or *dpy-10⁻/dpy-10⁻* hermaphrodites from the self-fertilization based on the phenotypes. You will have to pick several hermaphrodites because half of the worms will be doubly heterozygous and the other half will be homozygous for one gene and heterozygous for the other gene. Because only half will have the genotype required to produce a doubly homozygous worm, there will be (1/2)(1/4)(1/4) or 1/32 of the worms will have the genotype desired. Almost all of the progeny will be hermaphrodites because the last step is a selfing.

20-10. a) You use the wild-type males (who are heterozygous for *unc-54⁻*) and cross to the *daf-8⁻/daf-8⁻* hermaphrodites.

b) The progeny from the hermaphrodite self-fertilization are *daf-8⁻/daf-8⁻*. The progeny from the outcross are

Males:

$$\frac{unc54^+\ daf\text{-}8^+}{unc54^+\ daf\text{-}8^-} \quad \text{and} \quad \frac{unc54^-\ daf\text{-}8^+}{unc54^+\ daf8^-}$$

Hermaphrodites:

$$\frac{unc54^+\ daf\text{-}8^+}{unc54^+\ daf\text{-}8^-} \quad \text{and} \quad \frac{unc54^-\ daf\text{-}8^+}{unc54^+\ daf\text{-}8^-}$$

All are wild-type.

c) The wild-type hermaphrodites will be selfed to get the double homozygotes (but half of these worms will be heterozygous and the other homozygous for *unc-54⁺*. The *unc-54⁺* homozygotes will not produce double homozygotes.)

d) Because the *daf-8* and *unc-54* genes are linked, there has to be recombination (crossing-over) between these genes during meiosis to generate gametes containing *unc-54⁻* and *daf-8⁻* in the same gamete.

e) The genes *daf-8* and *unc-54* are 18 map units apart, so 18% of the <u>gametes</u> will be recombinant and 9% will contain the *unc-54⁻ daf8⁻* combination of alleles. The probability of two gametes with this genotype combining is (.09)(.09). Because only one-half of the wild-type hermaphrodites are the double heterozygotes, 1/2 (.09)(.09) or .004 or 0.4% of the progeny will have the desired double homozygous genotype.

20-11. a) The candidate gene approach would involve testing the suspected gene (previously identified K⁺ channel gene) in the mutant worm to determine if the defect in that worm is in the channel gene. The candidate gene would be cloned from the mutant and the sequence determined to look for a mutation in that particular K⁺ channel gene. A change in sequence in the mutant is suggestive that the suspected gene is involved. If there were a number of independent alleles, all of which showed a change in the sequence of the gene, that would be much better evidence. Stronger evidence would come from introducing a wild-type copy of the cloned candidate gene into worms by transformation and look for rescue of the phenotype. Another alternative is to use the RNA interference approach. Inject worms with double stranded RNA (made in vitro) corresponding to the gene and see if this results in a phenotype similar to that of the mutant.

b) You could determine if the Tc1 insertion occurred in the ion channel gene by Southern hybridization analysis. The probe would be a normal copy of the K⁺ channel gene and you would look for an alteration in the restriction pattern (an addition of DNA equivalent to the size of the Tc1 element) of the channel gene compared to wild-type. Alternatively, you could use PCR with one primer consisting of a Tc1 sequence and the other primer being a sequence from the ion channel gene. If amplification occurs and a fragment is produced, the Tc1 element must be in the vicinity of the ion channel gene.

c) The mutation would probably have a dominant phenotype because it does not result in lack of a channel but produces a channel with an altered function. In the heterozygote, it is likely that some of the channels in each cell would remain open inappropriately if those channels were constructed from the mutant protein.

d) You could compare the phenotypic response to isoamyl alcohol and diacetyl alcohol in the wild-type and *egl-2*⁻ mutant strains. (Attraction or repulsion to these two substances is the phenotype that would be tested.)

20-12. a) The mouse netrin gene could be cloned and introduced into an *unc-6*⁻/*unc-6*⁻ mutant worm to determine if the mouse gene could substitute for the worm gene and correct the uncoordinated phenotype. The coding region of the mouse gene should be cloned downstream of a copy of the promoter of the worm gene to ensure that the mouse gene would be transcribed in the right tissues and in the right amounts.

b) The migration of other cells could be compared in the *unc* mutants and wild-type worms.

c) You could identify interacting proteins by isolating suppressors of the *unc-5*⁻ mutations. The new mutations, in interacting genes, compensate for the defect in the *unc-5*⁻ mutants.

d) You would clone the gene next to a promoter that was expressed at all times in all cells (a housekeeping type of gene, for example).

Chapter 21   *Drosophila melanogaster:* Genetic Portrait of the Fruit Fly

Synopsis

     Drosophila has been a very important model system for studying chromosome structure and development because of the ease of doing genetic and cytogenetic analyses in this organism. Our understanding of pattern development, an underlying theme of development, is based on the pioneering genetic studies in *Drosophila*. Many features of *Drosophila* biology have been used and further developed as research tools.

- As in other model organisms, the life cycle is rapid, allowing many generations to be studied in a short amount of time.

- Banding patterns in polytene chromosomes allow the gross-level mapping of chromosomes and chromosomal rearrangements.

- Small genome size and small number of chromosomes (4) makes cytological and mapping experiments easier.

- Balancer chromosomes, having multiple inversions, a dominant genetic marker and a recessive lethal allele allow the maintenance of a recessive lethal mutation on the other homolog.

- P element mutagenesis creates mutations that are tagged with a transposon, making cloning of the gene easier.

- Enhancer traps can be used to identify genes with specific expression patterns (e,g., in specific tissues or at certain times in development).

- Mosaic flies can be created during development by X-ray induced mitotic recombination.

- Ectopic expression of a gene (expression in a different cell or at a different time than normally occurs) can be created by introducing a transgene linked to a promoter that is expressed in the desired way.

Be prepared to:

After reading the chapter and thinking about the concepts, you should be able to:

- set up crosses using balancer chromosomes

- create maps by crossing deletions and point mutations

- analyze results of in situ hybridization of probes to mutant and wild-type embyros

- distinguish between maternal effect mutations and zygotic mutations
- design experiments to look for homologous genes in other organisms and test the functionality of those genes in *Drosophila*
- think about how genetic analysis is done in this organism including: isolation of mutants, characterization of mutations (complementation and epistasis analyses), mapping a gene, cloning a gene, making a mutant cloned copy and reintroducing it into the genome

Problem-solving tips:

- Draw out chromosomes in parents and in their gametes to determine the genotypes and phenotypes of progeny expected.
- In balancer chromosomes, the dominant mutation is a marker that indicates which flies received the balancer; the recessive lethal mutation means that a fly receiving two copies of the balancer will not live; the multiple inversions prevent crossing-over between the balancer and a normal chromosome.
- P flies contain P elements; M flies lack P elements. Introduction of a P element into an M fly results in transposition of the P element in the germline of the recipient.
- Enhancer trapping is used to search for regulatory regions whose patterns of expression are recognized by expression of the *lacZ* gene in the introduced fusion.
- A gene product is cell autonomous if it affects only the cell in which it is made.
- Maternal effect mutations are not expressed in the female carrying the mutation, but are expressed in her progeny. If the mother is homozygous for a maternal effect mutation, she is not affected but all her progeny will be, regardless of their genotype.
- Some of the longer problems in this chapter are real research problems that were carried out essentially as described to you. They require that you integrate your knowledge of classical genetics, molecular genetics, genome analysis. These problems give you more practice in thinking like a geneticist. Think about how you would discover the genes involved and how their gene products may act- as regulators, messengers, structural components, etc.

Solutions to problems

21-1. a.4 b.5 c.1 d.6 e.3 f.2

21-2. a) Complementation is measured because the two mutations are present in the progeny being examined but there has been no chance for meiotic recombination yet.

b) In the cross between mutant 1 and mutant 2, there are four genotypic classes of progeny: m1/m2; m1/TM3; m2/TM3; TM3/TM3. The first class will be mutant if the mutations are in the same gene but will be wild-type if the mutations are in different genes and therefore complement one another. The second and third genotypic classes will have the Stubble phenotype because of the presence of one copy of the balancer. The fourth class will be dead because of the recessive lethal phenotype of Stubble. The 103 progeny that are not wild-type in this cross are progeny of the second and third genotypic classes and have the Stubble phenotype.

c) The 50 zygotes that do not live are homozygous for the balancer, TM3/TM3.

d) There are 6 genes represented by these 10 mutations. Mutations in the same gene do not complement each other and therefore there will not be any wild-type progeny produced. Complementation groups are 10,1; 9,5; 7,8,4; 6; 3; 2.

e) complementation

f) If no wild-type progeny were produced, there was no complementation between the point mutation and the deletion mutation when chromosomes containing each of these mutations are present in the progeny. Therefore the mutation maps within the deleted area. Convert the information in the table into a different form. For each deletion, the following mutations map within the deleted area:

Deletion	Genes mapping within the deletion
A	2, 4, 6, 7, 8
B	1, 2, 3, 4, 6, 7, 8, 10
C	1, 3, 5, 9, 10
D	3, 4, 6, 7, 8
E	1, 3, 4, 7, 8

To determine the order, you can start by looking for a common set of deleted genes in different mutants (for example, 4, 6, 7, 8), then compare the additional genes deleted. Comparing delA and delD, gene 2 is missing in delA; gene 3 is missing in delD. Therefore genes 2 and 3 flank the group 4, 6, 7, 8 on opposite sides. A comparison of delD and delE indicates that gene 6 is not next to gene 3 but would be next to gene 2. Use the remaining deletions to place the other genes using similar logic. DelC removes gene 3 and tells us that the additional genes 1, 5, 9, 10 are next to gene 3. Comparison of delE and delC shows that 1 and 10 are next to 3. On the map below, the parentheses indicate genes for which the order cannot be determined with the information given.

```
_____A_____
_____B_____
                    _____C_____
          _____D_____
              _____E_____
  2    6    (4,7,8)   3   1 10   (5,9)
```

21-3. There are several possible ways to do this- we outline one. You could start with male flies that had one or more insertions of the *w+* marked P elements on the X chromosome. These male flies must have a source of transposase elsewhere in the genome. There will be P element transpositions in the germline of these males. You now mate these males to females who are homozygous for *w-* mutations on their X chromosomes. In the next generation, you look for male flies that have pigmented eyes. These flies must have gotten their X chromosome from their mother, so the pigmentation must have arisen from transposition in the father's germline that created sperm with the *w+* P element on an autosome or on the Y chromosome. To distinguish the autosomal inserts versus the Y chromosome inserts, you take these pigmented males and again mate them to *w-/w-* females. If the P insert is autosomal, some males and some females will have pigmented eyes. If the P insert is on the Y, all the males but none of the females will have pigmented eyes.

21-4. c. P element transposition will occur in the germline of the F_1 progeny of a male containing a P element (P male) crossed to a female lacking P elements (M female).

21-5. e.

21-6. a) The promoter sequence itself is recognized by RNA polymerase. The region also contains binding sites for transcription factors such as Bicoid that ensure that the *hb* gene is transcribed only in zygotic nuclei in the anterior part of the egg and binding sites for other transcription factors that ensure the *hb* gene is transcribed in the proper cells in the mother so that it is deposited uniformly in the egg before fertilization.

b) The coding region contains information for the DNA binding domains and domains involved in the transcriptional regulation of gap and pair rule genes.

c) The sequence could be needed for translational repression (carried out by Nanos protein), thereby preventing translation of the maternally supplied *hb* mRNA in the posterior portion of the embryo.

21-7. Segment polarity genes encode proteins involved in signal transduction while gap and pair rule genes encode transcription factors. The segment polarity genes act after cellularization in the embryo whereas the gap and pair rule genes act before cellularization so the gene products can freely diffuse in the syncytium in the embryo.

21-8. The anterior cytoplasm from a wild-type embryo could be injected into the anterior end of a *bicoid* mutant embryo to see if there was rescue of the mutant phenotype. You would want a control that would indicate that the physical act of injection does not cause rescue. The control would be injection from a *bicoid* mutant embryo into a mutant embryo. Purified *bicoid* mRNA injected into the anterior end of a *bicoid* mutant embryo would be a more definitive experiment indicating that *bicoid* alone is sufficient to rescue. Finally, purified *bicoid* mRNA could be injected into the posterior end of a wild-type embryo. If *bicoid* is an anterior determinant, there should be two anterior ends developing.

21-9. These results indicate that 1) maternally-supplied *hunchback* mRNA is completely dispensable and 2) that the function of the Nanos protein is only needed to restrict the translation of the maternally-supplied *hunchback* mRNA in the posterior of the egg. Development is fine if there is no maternally-supplied *hunchback* mRNA, so if there is no *hunchback* mRNA in an embryo, Nanos is not needed and can be defective (mutant) without showing any effect. If there is too much maternally-supplied *hunchback* mRNA, this swamps out the Nanos protein supplied even by wild-type mothers. The result is too much Hunchback protein in the posterior of the egg and this prevents abdominal development. This is a peculiar situation : the fly doesn't really need *hunchback* maternal mRNA at all, but it makes it anyway. Because it makes this unnecessary *hunchback* maternal mRNA, it now needs *nanos* and the other posterior group genes to prevent its translation in the posterior of the egg. This illustrates that evolution does not always come up with the simplest solution to a problem, just one that happens to work. You should note that Hunchback is still needed for proper development even if the maternal hunchback mRNA is not; this is because *hunchback* must still be transcribed from zygotic nuclei so it can perform its role as a gap gene.

21-10. a) Knirps has no effect on *hb* protein distribution but does affect Kruppel localization. More specifically, Knirps protein is needed to restrict the posterior limit of the zone of Kruppel expression. b) Hunchback protein would be seen throughout the embryo because there is no Nanos protein to inhibit its translation.

21-11. a) There is more than one progenitor cell that divides to form the cells within any one facet. b) These results show that the expression of the white gene is cell autonomous; the phenotype of any cell (red versus white) is dictated by the genotype of that cell for the white gene.

21-12. To solve the mapping, look first at the deletions. If no non-Cy progeny are obtained, there was no complementation and the deletions must have been overlapping. Deletion A overlaps B and deletion A overlaps C. But B and C do not overlap. The map position of the point mutations with respect to the deletions can be determined by looking for non-complementing lethal mutations or uncovering of a sterile phenotype when crossed to the deletion (because the mutation is the only copy present in the strain with a deletion on the other homolog).   Mutation 4 is uncovered as a sterile by combining with delA or delB, so the gene must be located in the region of overlap between those two deletions. Similarly mutation 1 fails to complement delA or delC so the gene is located in the region of overlap between these two deletions. Mutations 5 and 6 (which do not complement each other for fertility) are uncovered by deletion C only, and mutation 3 is uncovered by deletion A only. Mutations 2 and 7, which do not complement each other for viability and therefore must be in the same gene, are also not complemented by delB.

```
                  '      _____A_____
      _____B_____          _____C_____
      2,7            4       3       1        5, 6
```

b) five genes: 1; 2,7; 3; 4; 5,6

c) Three genes (3; 4; 5,6) are female steriles as indicated by the phenotype when combined with a deletion that covers that gene. If the mutant females (females homozygous for a female sterile mutation, or heterozygous for two different female sterile mutations in the same gene) make normal-looking unfertilized eggs that become fertilized but arrest their development soon afterward, the mutations are maternal effect lethal mutations. If the eggs laid by these females are abnormal or are unable to be fertilized, the genes disrupted by the female sterile mutations are involved in oogenesis.

d) Zygotic lethal genes are 1; 2,7 as indicated by the lack of viable non-Cy progeny in crosses with deletion strains.

e) You could not determine map distances between mutations in the same gene because the double mutant 2,7 do not survive and the double mutant 5,6 is sterile.

21-13. a) By obtaining a *lacZ* insertion into a developmental gene you could study the time of expression during development and the cells in which the gene is expressed. (This assumes that the *lacZ* gene is turned on at the same time and in the same place as the developmental gene into which it inserts. This appears to be true in most, but not all cases.)

b) The female flies would have to be heterozygous and could be maintained in a cell containing a balancer chromosome corresponding to the chromosomes on which the mutation was located.

c) You could compare the pattern of *bicoid* mRNA in *homeless* vs wild-type embryos by hybridization using a *bicoid* DNA probe versus mRNA from the two types of embryos (The control is the wild-type mRNA. An additional control would be to examine the localization of other mRNAs in the embryo such as the *nanos* message to make sure that the effect of the *homeless* mutation is specific to *bicoid*, and that it is not simply that the mutation results in the non-specific degradation of many RNAs in the embryo.

d) You would make females of the genotype   centromere - ovo^D - ho^+ / centromere - ovo^+ - ho^- . You would then induce mitotic crossingover in these females. The females would normally be unable to produce ovaries, but they could produce an ovary if mitotic crossingover <u>within the germline</u> yields a cell that is homozygous for the *homeless* mutation. If all females with such a germline clone are sterile, this implies that the *homeless* gene function is needed in the germline. If these females are fertile, it implies that *homeless* gene function can be supplied by non-germline cells (that is, in the somatic follicle cells.)

e) Ras is needed for several different pathways, but the level of Ras protein required is different in the various signaling pathways.

12-14. (Note that the *bruno* gene has been cloned, but that there are no known mutants of this gene, while mutants of *arret* have been isolated, but the gene has not yet been cloned.)

The simplest test would be to clone the wildtype bruno gene, and to insert it into a vector for P element-mediated transformation. This would be transformed into flies, and then crosses would be made to establish whether the $bruno^+$ transgene can rescue the fertility of *arret* homozygotes. If this were the case, you would then ask whether the *oskar* mRNA was properly localized in eggs produced by *arret* mutant mothers. As a supplementary experiment, you could clone the *bruno* gene from *arret* mutants, and sequence these genes to determine if there are potentially significant mutations in the *bruno* gene in *arret* mutants.

Chapter 22   *Mus musculus:* A Genetic Portrait of the House Mouse

Synopsis:

The most significant techniques developed for analysis of genes and their function in the mouse are the transgenic technologies. These techniques and other characteristics listed below have increased out abilities to use the mouse as a model for human disease as well as study developmental questions.

- similarity to humans in genetic organization (conserved synteny in genetic maps of the chromosomes)
- similarity of physiology means that models for human diseases can often be created in the mouse
- creation of transgenic mice in which a piece of foreign DNA is injected into the nucleus of an egg, where it inserts into the genome
- targeted mutagenesis using ES cells

Be prepared to:

After reading the chapter and thinking about the concepts, you should be able to:

- understand the consequences of transgenic technology and targeted mutagenesis – what happens within the genome
- think through crosses from generation to generation
- analyze patterns of mRNA and protein expression and describe that regulation or type of mutation based on these patterns
- think about how genetic analysis is done in this organism including: isolation of mutants, characterization of mutations (complementation and epistasis analyses), mapping a gene, cloning a gene, making a mutant cloned copy and reintroducing it into the genome

Problem-solving tips:

- Transgenic technology can result in an additional copy of the wild-type gene or introduction of an extra copy that is a mutant copy (add-on type of analysis). Another type of transgenic technology is targeted mutagenesis in which a mutated copy of a gene replaces a normal copy.
- After replacement of one copy of a gene with a mutated copy, mice containing the replacement on one copy of the homolog must be mated to obtain a homozygous knockout mutant mouse and study the gene's function.

- Analysis of patterns of expression using a reporter gene such as *lacZ* can give clues as to the function of a gene.

- FISH (fluorescent in situ hybridization) allows one to quickly locate the chromosome on which a gene is located.

- Some of the longer problems in this chapter are real research problems that were carried out essentially as described to you. They require that you integrate your knowledge of classical genetics, molecular genetics, and genome analysis. These problems give you practice in thinking like a geneticist. Think about how you would discover the genes involved and how their gene products may act- as regulators, messengers, structural components, etc.

Solutions to Problems

22-1. a.3 b.5 c.1 d.6 e.7 f.2 g.4

22-2. a) Because the genome is so large, many more clones have to be screened to find a single copy gene of interest. The large genome size is due to a large amount of repetitive DNA. This can create problems for finding a piece of unique DNA to use for finding a specific clone and can create some problems for mapping clones within the genome.
b) Mouse is useful as a mammalian model system for human diseases and human specific physiology.

22-3. a) yeast. Because this is a general question about Cl⁻ channel function, it could be addressed in yeast where the creation of specific mutants is easier and quicker.
b) mouse. Because the question is specifically addressing a mammalian disease and mammalian physiology, the question should be studied in mice.
c) yeast. In yeast the genetic experiments and gene replacements are much easier to do and quicker to analyze.
d) yeast (and mouse) You might want to start the drug discovery process in yeast where it is easy to test several drugs quickly, but after finding good candidates, you could use the mouse to determine if the new drugs had any other ill effects in mammals.

22-4. a) FISH – fluorescent in situ hybridization (see chapter 10) using a piece of the mouse gene as a probe versus a human chromosome spread. (The alternative of finding linked molecular markers is much more time and resource intensive.)

b) Use the mouse gene as a probe against a human genomic library or a library specifically containing chromosome 8 clones. The clone may contain other genes as well. The fragment(s) containing the gene could be identified by further hybridizations to blots of the restriction digested clone.

22-5. The fertilized egg in mammals can split into two separate cells or two separate parts during the first divisions, but in *Drosophila,* because the zygote is a syncytium with a polarity, division of the embryo in two will result in loss of important information.

22-6. The ES cells contain a different coat color allele than the cells of the blastocyst with which they will be combined. The chimeric embryo that develops will have some cells of one coat color and other cells of the other coat color. This makes it easy to recognize those mice that developed from chimeric embryos.

22-7. a) Because there are no -/- mice, the *CTF* gene seems to be an essential gene.

b) The *CTF* gene affects growth in both males and females and is necessary for fertility in males.

c) A disrupted allele generally produces a different restriction fragment pattern in that portion of the genome when analyzed by Southern analysis. A small tissue sample (blood) would be used to determine if the mice were homozygous or heterozygous.

22-8. a) Mutant A must be a small point mutation. Mutant B is defective in making protein or an aberrant protein is made and degraded. In mutant C there is a truncated RNA, but a normal DNA, so there may be an altered RNA transcription or processing signal in the mutant. In mutant D, the DNA and RNA are both altered, suggesting that there is a deletion in the DNA. In mutant E, no RNA or protein are made and the DNA is altered, so the regulatory region (transcription start) may be deleted. Mutant F appears to be a deletion of the entire gene.

b) A cross could be done using a marker linked to the RAR locus to see if the defect cosegregates with the marker. If the defect was in a different gene, it would not cosegregate with the marker.

c) You could clone the RAR gene next to a neuron cell specific promoter and create a transgenic mouse that now expressed RAR in neurons.

22-9. a) If the mutation was due to an insertion of the transgene, the MMTV *c-myc* gene should segregate with the phenotype. That is, all subsequent animals that had the limb deformity should have the *c-myc* fusion and vice versa. The presence of the *c-myc* fusion could be recognized by Southern analysis.

b) Clones containing the *c-myc* fusion could be identified by hybridization of MMTV sequences versus a library of genomic clones produced from the cells of the mutant mouse. The DNA surrounding the MMTV *c-myc* fusion in this clone would be the gene of interest.

c) The sequence of the gene into which the transgene inserted could be analyzed in the *ld* mutant to determine if there were mutations in the gene. Alternatively a clone containing only the wild-type copy of the gene into which this transgene inserted could be be injected into a mouse heterozygous for the *ld* mutation. If the homozygous *ld/ld* progeny that had also received the transgene are not wild-type, the transgene is not the *ld* gene.

d) The presence of transcripts in both embryo and adult is still consistent with role in development. The expectation would be transcription in the embryo. The presence of transcripts in adults may indicate that the gene also plays a role in adults as well.

22-10. a) The data are consistent with the hypothesis because the level of Ob protein is low in the animals that are starved and highest in those that were force-fed the high calorie diet.

b) To determine if the Ob gene is transcriptionally regulated, the Ob DNA would be used as a probe vs RNA from starved and normal (or force-fed) mice. If regulation occurs at the transcriptional level, you would expect the lowest levels in the starved and high levels in force-fed mice.

c) If the mRNA levels are the same (while protein levels are different), the regulation must be at the translational level.

d) Mutant A has the normal sized RNA, so the mutation must not be a large deletion or alteration in the RNA processing. Mutant B has no RNA, so could be a deletion of the entire gene or a defect in the regulatory region that knocks out the transcription start. Mutant C has a larger than wild-type transcript so could have a defect in processing (e.g., does not splice out an intron) or a deletion of DNA that fused the *Ob* gene to another gene producing one large hybrid transcript. Mutant D has a smaller than normal transcript which could mean there is a deletion in the gene or an altered processing (splicing) signal.

e) 1.B (Fat cells are the cells where you expect the gene would be expressed.) 2.A (You want to express the gene in mouse cells, so should use a mouse promoter you can control.) 3.A. (Addition of metal (Zn) will induce expression of the fusion gene.)

f) Obese. Lacking the Ob receptor, mice would not be able to receive the signal from the Ob protein that they were full and probably would continue to eat and become obese.

g) Injection of the Ob protein into mice where Ob is not normally made could rescue the phenotype- that is the mice might not continue to eat large quantities of food and become obese. The regulation of the protein levels might be critical- so the mice don't stop eating when they need food.

Chapter 23   The Genetic Analysis of Populations and How They Evolve

Synopsis

   This chapter involves the study of how genetic laws impact the genetic makeup of a population. Mendelian principles are the basis for the Hardy-Weinberg law which allows one to calculate allele and genotype frequencies from one generation to the next. The Hardy-Weinberg law can be used only if other forces are not acting on the allele frequency. Those forces include selection, migration, mutation, and population size.

   Population geneticists try to determine the extent to which a trait is determined by genetic factors and how much is determined by environmental factors. Knowing if a trait is largely determined by genetic factors introduces the possibility for animal and plant breeders to select and maintain populations with desired traits.

Be prepared to:

After reading the chapter and thinking about the concepts, you should be able to:

- determine allele frequencies in a population given the frequencies of genotypes
- determine genotype frequencies in a population given the frequencies of alleles
- determine genotype and/or allele frequencies in the next generation given the genotype or allele frequencies in the present generation
- determine if a population is in equilibrium
- determine allele and genotype frequencies after migration has occurred
- describe how heritability of a trait can be determined

Problem-solving tips:

- p, q are representations of allele frequencies.
- p^2, 2pq, q^2 are representations of genotype frequencies.
- Once a population is at equilibrium, allele and genotype frequencies do not change.
- If a genotype is selected against or if populations are combined (by migrations) or if there is significant mutation (usually together with selection), allele frequencies will change.
- selection for one genotype and selection against another genotype are balanced at a particular allele frequency (equilibrium frequency).
- Genetic drift is most often seen in small populations.

- Polygenic traits are controlled solely by alleles of two or more gene; multifactorial traits include polygenic traits and traits that are influenced by both genes and environment.
- Genetic and environmental contributions to a phenotype are sorted out by setting up conditions in which the genetic background is consistent (to analyze environmental contributions) or conditions in which the environment is constant (to analyze genetic contributions).

Solutions to Problems

23-1. a.3 b.5 c.8 d.7 e.6 f.1 g.9 h.2 i.4

23-2. a) To calculate genotype frequencies, divide the number of frogs with each genotype by the total number of frogs.

G^GG^G=120/200 or 0.6

G^GG^B=60/200 or 0.3

G^BG^B=20/120 or 0.1

b) The allele frequencies are determined by totalling all alleles within each genotype.

G^GG^G 120 individuals with 2 G^G alleles = 240 G^G alleles

G^GG^B 60 individuals with one G^G allele = 60 G^G alleles

G^GG^B 60 individuals with one G^B allele = 60 G^B alleles

G^BG^B 20 individuals with two G^B alleles = 40 G^B alleles

There are 300 G^G alleles/400 total alleles so the frequency (p) of G^G is 0.75.

There are 100 G^B alleles/400 total alleles so the frequency (q) of G^B is 0.25.

c) The expected frequencies can be calculated using the allele frequency and the terms of the Hardy-Weinberg law.

G^GG^G=p^2=0.5625

G^GG^B=2pq= 0.375

G^BG^B=q^2= .0625

23-3. a) The allele frequencies are calculated from the proportion of individuals with different genotypes. The problem becomes easier and more intuitive if you assume the population has 100 individuals.

0.5 MM individuals = 50 individuals	= 100 M alleles
0.2 MN individuals = 20 individuals	= 20 M alleles
	= 20 N alleles
0.3 NN individuals = 30 individuals	= 60 N alleles

120M alleles/ 200 total alleles = 0.6 - allele frequency for M

80 N alleles/200 total alleles = 0.4 - allele frequency for N

The expected genotype frequencies in the next generation are calculated using these allele frequencies.

MM= p^2 = 0.36

MN= 2pq = 0.48

NN = q^2 = 0.16

b) Non-random mating, small population size, migration into the population, mutation, or selection could cause a departure from the Hardy-Weinberg equilibrium.

23-4. a) There are 60 flies with normal wings (+/+) out of 150 total flies and 90 flies with the heterozygous genotype.

60 (+/+) flies	= 120 + alleles
90 (+/Delta) flies	= 90 + alleles
	= 90 Delta alleles

120 + 90 "+" alleles = 210/300 or allele frequency of 0.7

90 Delta alleles/300 = allele frequency of 0.3

b) In the second generation, the following offspring would be produced:

p^2= $(0.7)^2$ = 0.49

2pq= 2(0.7)(0.3) = 0.42

q^2 = $(0.3)^2$ = .09

The homozygous q^2 (Delta/Delta) do not live, so, the 0.49 and 0.42 remaining makeup a total of 0.91

0.49/0.91 = 0.54 = proportion of the viable offspring with the +/+ genotype

0.42/0.91 = 0.46 = proportion of the viable offspring with the +/- genotype

The numbers of individuals with these two genotypes in the viable population would be 0.54 × 160 = 86.4 (+/+) and 0.46 × 160 = 73.6 (+/Delta)

23-5. For each of the populations, determine the allele frequency using the genotype frequencies in the population, then calculate the expected genotype frequency for that allele frequency when a population is in equilibrium.

a) yes, population a is in equilibrium.

0.25 AA   = 50 A alleles

0.50 Aa   = 50 A alleles

                  = 50 a alleles

0.25 aa   = 50 a alleles.

100A alleles/200 = 0.5 = p

100 a alleles/200 = 0.5 = q

p^2 =0.25; 2pq = 0.5; q^2 = 0.25.

These genotype frequency numbers match those seen in the population, so it is at equilibrium.

b) no, population b is not in equilibrium

0.1 AA   = 20 A alleles

0.74 Aa   = 74 A alleles

                  = 74 a alleles

0.16 aa   = 32 a alleles

94 A alleles/200 alleles = 0.47 = p

106 a alleles/200 alleles = 0.53 = q

p^2 = 0.22; 2pq = 0.5; q^2 = 0.28 (Numbers do not match that seen in the population.)

c) no, population c is not in equilibrium

0.64 AA   =128 A alleles

0.27 Aa   = 27 A alleles

                         = 27 a alleles

0.09 aa   = 18 a alleles

128 +27 = 155/200 alleles = 0.78 = p

27 + 18 = 45/200 alleles = 0.22 = q

p^2= 0.61; 2pq= 0.34; q^2= 0.05 (Numbers do not match the frequencies seen in the population.)

d) no, population d is not in equilibrium

0.46AA   = 92 A alleles

0.50 Aa   = 50 A alleles

                        = 50 a alleles

0.04 aa   = 8 a alleles

92 +50 A alleles/200 = 142/200 = 0.71 = p

58 a alleles/200 = 0.29 = q

p^2=0.50; 2pq = 0.41; q^2= 0.08 (Numbers do not match those seen in the population.)

e) yes, population e is in equilibrium

0.81 AA	=162 A alleles
0.18 Aa	= 18 A alleles
	= 18 a alleles
0.01 aa	= 2 a alleles

162 + 18 A alleles/200 = 180/200= 0.9 = p

18 + 2 a alleles/200 =0.1 = q

p^2 = 0.81; 2pq = 0.18; q^2 = 0.01 (Numbers do match the genotype frequencies seen in the population.)

23-6. Consider the Q and R genes separately. In each case, you need to determine the frequency of each allele, then calculate the genotype frequencies expected if the population is in equilibrium for that gene.

The number of Q^F alleles in the population is 202 + 202 + 101 + 372 + 372 + 186 + 166 + 166 + 83 = 1850.

The number of Q^G alleles is 101 + 101 + 101 +186 + 186 + 186 +83 + 83 + 83 = 1110.

The total number of alleles is 2960.

The frequency of the Q^F allele is 1850/2960 = 0.625.

The frequency of the Q^G allele is 1110/2960 = 0.375.

From these numbers you can calculate the expected genotype frequency. p^2= 0.39; 2pq = 0.47; q^2 = 0.14.

The genotype frequencies in the population are the following.

For Q^FQ^F, 202 + 372 + 166 = 740/1480 = 0.5. For Q^FQ^G, 101 +186 + 83 = 370/1480 = 0.25. For Q^GQ^G, 101 + 186 + 83 or 370/1480 = 0.25.

Since the expected and observed are not similar, the population is not in equilibrium for the Q gene.

The number of R^C alleles is 202+ 202 + 101 + 101 + 101 + 101 + 372 + 186 + 186 = 1552.

The number of R^D alleles is 372 + 186 + 186 + 166 + 166 + 83 + 83 +83 + 83 = 1408.

The frequency of the R^C allele is 1552/2960 = 0.52.

The frequency of the R^D allele is 1408/2960 = 0.48.

The expected genotype frequencies for these allele frequencies are p^2= 0.27; 2pq = 0.5; q^2= 0.23.

The observed genotype frequencies are the following. For R^CR^C: 202 + 101 + 101 = 404/1480 = 0.27; for R^CR^D: 372 + 186 + 186 = 744/1480 = 0.5; for R^DR^D: 166 + 83 + 83 = 332/1480 = 0.22. These numbers are very close to that expected, so the population is in equilibrium for the R gene.

b) In the next generation, with random mating, the fraction that will be Q^FQ^F will be that which we calculated as the expected genotype frequency in part a: 0.39 for the Q^FQ^F genotype.

c) The fraction that will be R^CR^C in the next generation will again be the expected frequency calculated based on the allele frequency: 0.27.

d) The prediction on the genotype of a child of two specific parents in the population is not influenced by allele frequency. This is a standard probability question starting with specific parents $Q^FQ^GR^CR^D$ and $Q^FQ^FR^CR^D$. The probability is the sum of the individual probabilities for each of the genes. There is a 1/2 chance that the female will contribute the Q^G allele and a 1/1 chance that the male will contribute the Q^F allele. There is a 1/4 chance that the parents, heterozygous for the R^C and R^D allele, will be homozygous. There is a 1/2 chance the child will be male. The overall probability is $(1/2)(1/2)(1/4) = 1/16$.

23-7. Each different allele frequency has a different set of genotype frequencies at equilibrium. There is not one equilibrium point that an allele will go to.

23-8. The 3:1 ration is seen when two heterozygous **individuals** are crossed- this ratio is not relevant for a <u>population</u>. The ratio will depend on allele frequency.

23-9. a) For the sailor population:

324 MM sailors	648 M alleles
72 MN sailors	72 M alleles
	72 N alleles
4 NN sailors	8 N alleles

$72 + 8 = 80$ N alleles/800 total alleles $= 0.1$

b) When the population is mixed together (random mating), the allele frequency in the mixed population can be calculated and the genotype frequency after random mating determined.
You need to determine the genotype frequency and use that to determine the number of Polynesians with each genotype to be able to determine the number of alleles they contribute to the pool.
Polynesian allele frequency p=.06, q=.94.

p^2 (MM) = .0036; 600 individuals $\times$ .0036 = 2 MM individuals

$2pq$ (MN) = .1128 600 individuals $\times$ .1128 = 68 MN individuals

q^2 (NN) = .8836 600 individuals $\times$ .8836 = 530 NN individuals

Now determine the allele frequency in the mixed population.

	sailor	Polynesian
MM	648 M alleles	4 M alleles
MN	72 M alleles	68 M alleles
	72 N alleles	68 N alleles
NN	8 N alleles	1060 N alleles

M alleles = 648 + 4 + 72 + 68 = 792/2000 = 0.396

N alleles = 72 + 68 +8 + 1060= 1208/2000 = 0.604

$2pq$ (MN) frequency will be .478. The number of children with the MN genotype will be .478 x 1000 or 478.

c)

genotype	number of children	number of alleles
MM	50	100M
MN	850	850M
		850N
NN	100	200

The frequency of the N allele is 200/2000 or 0.1.

23-10. Evaluate the information you are given.

In the French population, individuals with the Ugh phenotype (could be either *AA* or *Aa* because Ugh is dominant) number 90. But there is only 50% penetrance, so twice as many individuals have those genotypes as the number that express the phenotype. That means that 180 individuals have the *AA* or *Aa* genotype and 320 have the *aa* genotype. Because the allele is in equilibrium, you know that $q^2 =$ *aa* frequency. q^2 therefore is 320/500 or 0.64. q = 0.8 and p =1-q or 0.2.

In the Kenyan population, 75 individuals had the Ugh phenotype and therefore were homozygous *AA* or heterozygous *Aa*. 150 individuals must have the *AA* or *Aa* genotype and the remaining 50 individuals had the *aa* genotype 50/200 or 0.25 = q^2 q=0.5 and p=0.5.

To determine the number of *A* and *a* alleles in the mixed population, you need to calculate the number of individuals with each genotype, then tally up the alleles.   For the French population,

p=0.2; q=0.8

p^2 *(AA)* = 0.04	0.04 × 500 = 20
$2pq$ *(Aa)* = 2(0.2)(0.8) = 0.32	0.32 × 500 individuals = 160 individuals
q^2 *(aa)* =0.64	0.64 × 500 individuals = 320 individuals

For the Kenyan population,

p=0.5; q=0.5

p^2=0.25	0.25 × 200 = 50 individuals
2pq=0.5	0.5 × 200 = 100 individuals
q^2=0.25	0.25 × 200 = 50 individuals

Calculating the number of alleles in the mixed population:

AA	20 × 2 = 40 A	50 × 2 =100 A	140
Aa	160 A	100 A	260
	160 a	100 a	260
aa	320 × 2 = 640 a	50 × 2 = 100 a	740

In the mixed population:

140 +260 = 400/1400 = 0.286 = p

260 + 740 = 1000/1400 = 0.714 = q

The number of individuals with the Ugh phenotype will be one-half of the individuals with the *AA* (p^2) and *Aa* (2pq) genotypes.

p^2= $(0.286)^2$ = 0.082; 0.082 (1000) = 82

2pq = 2 (0.286)(0.714) =0.408; 0.408(1000) = 408

(1/2)(408 + 82) = 245

The total number of Ugh individuals is 245.

23-11. a) 50% of the population shows high levels (over a broad range), so they must have either the 1/1 or 1/2 genotype. 50% have low levels so must have the 2/2 genotype. That means q^2 =0.50, q=0.71 and p (frequency of allele 1) = 1-q or 0.29.

b) In the babies born in the following year, the allele frequency will be the same- 0.29.

c) In the population of babies, use the p and q allele frequencies to determine genotype frequencies.

p=0.29; q=0.71

p^2 = 0.084

2pq = 0.412

q^2 = 0.50

The last genotype is selected against. In a population of 1000 individuals, there are 500 babies with this genotypes. 10% of the population is attacked by the virus, and 80% of those infected will die. Therefore, 50 will be infected and 40 of these will die. 460 with genotype 2/2 live.

The numbers after childhood will be:

460 with genotypes 2/2; 84 with 1/1; 412 with genotype 1/2. For allele 1, there are $84 \times 2 + 412$ or 580/1920 =.30

23-12. a) There are 150 mice.

60 t^+/t^+	120 t^+ alleles
90 t^+/t	90 t^+ alleles
	90 t alleles

210 t^+ alleles/300 total alleles = 0.7 allele frequency (p)

q = 0.3

b) First determine the frequencies of the three genotypes if all lived, then remove the inviable mice from your calculations and recalculate the allele and genotype frequencies.

$p^2 = 0.49$	49 individuals
$2pq = 0.42$	42 individuals
$q^2 = 0.09$	9 individuals

49 +42 survive = 91 total

49/91 are t^+/t^+ = 0.54

42/91 are t^+/t = 0.46

Of 200 individuals, 0.54 (200) or 108 are wild-type; 92 are tailless.

c) Determine the number of alleles contributed by each population to calculate the allele frequency for the mixed population then the genotype frequencies from the mating. Since the homozygotes die, you will have to subtract out these mice and recalculate the frequencies of the genotypes.

	Dom1	Dom2
t^+/t^+	$16 \times 2 = 32$ t^+ alleles	$48 \times 2 = 96$ t^+ alleles
$t^+/-$	48 t^+ alleles	36 t^+ alleles
	48 t alleles	36 t alleles

$32 + 96 + 48 + 36 = 212/296$ t^+ alleles = 0.716 = p

84 t alleles/296 = 0.284 = q

After mating, the following genotype frequencies are generated:

$p^2 = 0.512$

$2pq = 0.407$

$q^2 = 0.08$

But the t/t (q^2) mice die, so there are $512 + 407 = 919$ total living progeny (assuming 1000 progeny); $512/919 = 0.557$ = frequency of t^+/t^+ mice in the next generation and $407/919 = 0.443$ = frequency of t^+/t mice in the next generation

23-13. The frequency of colorblind males ($X^{cb}Y$) is equivalent to p since there is only one allele (one X) in these individuals. The frequency of colorblind females is p^2 ($X^{cb}X^{cb}$) = .0064.

23-14. *CF/CF* is 1/17,000 or $0.000059 = q^2$. q is 0.0077; therefore p = 0.9923. 2pq (frequency of heterozygote carriers) = 0.015 or 15 in 1000 (Note: In the Caucasian population the frequency of carriers is much higher – 1/20).

23-15. a) The F_1 flies are heterozygous and produced in the F_2 a population of flies consisting of 250 *Vg/Vg*, 500 *Vg/vg*, 250 *vg/vg*. The allele frequencies are 500 *Vg* + 500 *Vg* = 1000/2000 = 0.5 for each as they were in the F_1. When the *vg/vg* are dumped in the morgue the allele frequency in the mating population of F_2 flies changes so that there are now $250 \times 2 = 500 + 500 = 1000/1500$ or 0.666 *Vg* and 0.333 *vg*. That means that the genotype frequencies produced in the F_3 are $(0.666)^2 = 0.444$ for the *Vg/Vg*; 2(0.666)(0.333) = 0.444 for *Vg/vg* and 0.111 for *vg/vg*. The first two genotypes are wild-type so the frequency of wild-type flies is 0.888.
b) When the vestigial F_3 are dumped, the allele frequency shifts because the flies that are allowed to mate are 444 *Vg/Vg* and 444 *Vg/vg*. The frequency of the *Vg* allele is (2(444) + 444)/1776 or .75. The frequency of the *vg* allele is 0.25.
c) The genotype frequencies in the F_4 are calculated using the allele frequency of the F^3 generation. $p^2 = (0.75)^2$ or 0.5625; 2pq = 2(0.75)(0.25) or 0.375; $q^2 = (0.25)^2$ or 0.0625. The first two classes are wild-type, so 0.5625 + 0.375 = 0.9375. Now the vestigial winged flies are not dumped and random mating occurs. The frequency of wild-type and vestigial genotypes will be the same in F_5 as in F_4 - wild-type = 0.9375, vestigial = 0.0625.

23-16. The farther from equilibrium, the greater the Δq, so the population with an allele frequency of 0.2 will have the larger Δq.

23-17. A fully recessive allele is not expressed in a heterozygous organism, so there is no selection against the heterozygotes. The recessive allele is hidden in the heterozygote. In addition, a recessive allele sometimes confers an advantage when present in the heterozygote (as seen for the sickle cell allele in areas where malaria is prevalent). Finally, mutation can produce new recessive alleles in the population.

23-18. The equilibrium frequency will be different for the two populations. Equilibrium frequency is a balance between selection and mutation and the selection is very different in the two populations.

23-19. a) There is a founder effect of the descendants coming from a small number of individuals, so whatever recessive alleles were present in the population are more likely to be combined. Therefore the frequency of some alleles and genotypes can be higher in that population. Other alleles may not have been included in that original gene pool.

b) An advantage to studying the Finnish population is that there is genetic homogeneity and probably fewer genes (potential modifiers) that may affect the trait and therefore can be more easily dissected. A disadvantage is that some mutations that are present in general population may not be found in this small, inbred population and therefore will not be identified in studies of Finns.

23-20. a) Using genetic clones, only environmental effects contribute to variation.

b) Monozygotic twins are genetically identical so they can be thought of as genetic clones- compared to dizygotic twins who are genetically different. Comparison of MZ and DZ twins provides an assessment of the effect of genes versus environment.

c) Cross-fostering is removing offspring from a mother and placing with several different mothers to randomize the effects of different mothering environments. This is done to reduce environmental effects when determining heritability of a trait.

23-21. High heritability indicates that the phenotypic differences observed are due in large part to genetic differences. b) would be true.

23-22. a) $2n + 1$

b) A formula here would be $1/4^n$ There would be four pairs of genes controlling kernel color.

23-23. a) 32 cm represents the extreme phenotype, therefore the proportion of plants with this extreme phenotype are 1/44 or 1/256 or 0.0039. (When examining crosses between heterozygous individuals, you can think of the allele frequency as being effectively 0.5.)

b) When the frequency of the alleles in the population is not uniform as it is in a) of this question, you have to use the allele frequencies to determine the probability of obtaining a particular genotype. So, in the pool of gametes there is a (0.9)(0.9) probability that an AA genotype will form; a (0.9)(0.9) probability of a BB genotype; (0.1)(0.1) probability of CC genotype; and a (0.5)(0.5) probability of DD genotype. To get 32 cm all these allele combinations have to be present in one plant, so the overall probability is the product of all of these: 0.0016

Chapter 24   Evolution at the Molecular Level

Synopsis

This chapter takes a look at the evolution of the information of life, combining some of our recent knowledge of the structure and function of nucleic acids and the organization of information in the genomes with fossil record and paleontological evidence. Much of the information is speculative and hypotheses are proposed that are thought-provoking. Genome analysis provides information that drives the development of new hypotheses in this field.

Be prepared to:

After reading the chapter and thinking about the concepts, you should be able to:

- discuss the merits of DNA and RNA as information carrying molecules
- distinguish between synonymous and non-synonymous changes in DNA sequence
- describe how duplications can occur
- derive a molecular clock value and understand how to use it
- think creatively

Problem-solving tips

- Synonymous base changes in DNA do not change the amino acid that is encoded. Non-synonymous base mutations will result in a different amino acid being present in a to the protein.
- Duplications can occur by unequal crossing-over between repeated sequences or transposition of DNA.
- Molecular clock rates can be established by dividing the number of base changes by the length of time when the two organisms diverged (according to fossil or paleontological evidence).

Solutions to Problems

24-1. a.4 b.6 c.5 d.2 e.1 f.7 g.3

24-2. The fact that the same genetic code directs the cellular basis of life in varied organisms supports the unity of life hypothesis.

24-3. a. c.

24-4. a) RNA is a relatively unstable molecule, so would not be as good a storage information molecule as DNA. RNA can be degraded by chemical or enzymatic hydolysis (RNases). RNA is also not easily compacted.

b) Because RNA has only 4 "letters" in its alphabet, there is less variety than can be achieved in proteins.

24-5. a) The enzyme consists of an RNA molecule.

b) The enzyme has both an RNA and a protein component.

24-6. The differences should be genetic differences that affect many processes. The majority of the alterations should be effects on brain structure and function since the differences between humans and chimps are primarily in the cognitive functions. One way that many functions can be affected is by alterations in master regulatory genes (changing function or creating new regulators) that control the expression of many other genes.

24-7. a) The rates of nonsynonymous substitutions differ for these three genes because there is a different constraint on the function of each of the proteins. The function of the histones is very fixed in different organisms, so there is little room for variation, whereas growth hormone may have evolved to interact with other proteins that also have evolved.

b) The rates of synonymous substitutions are more constant between different genes because these base changes do not affect function of the gene product.

24-8. The immunoglobulin genes encode subunits of antibodies and a great diversity of antibodies is needed in the immune response. The changes that occur are therefore evolutionarily advantageous and would be maintained rather than being selected against.

24-9. The maintenance at a relatively high level suggests that there is some benefit to the *CF* allele in the heterozygous state. Recent evidence indicates that the allele may be somewhat protective against certain diseases. The *CF* allele may be similar to the sickle cell allele that protects heterozygous individuals against malaria.

24-10. A single copy of a gene can be duplicated if it is surrounded by small repetitive sequences and these sequences misalign during meiosis and crossing-over occurs. The gene that lies between the misaligned sequences will be duplicated.

24-11. Because the three color vision genes are cross-hybridizing, it is likely that they arose by duplication events and then diverged evolutionarily to function as different color receptors.

24-12. In transposition, a copy is transposed to a new location in the genome and therefore can be found at a very different location than the original copy. Unequal crossing-over produces extra copies adjacent to the original copy.

24-13. a) The A allele in humans and the M allele in mice are 600 base pairs different and mice and humans are 60 million years apart in evolutionary time. The clock rate we can derive from this information is that over 1 million years, 10 bp on average were changed. The A allele has 3000 base differences compared to *Xenopus* (X allele). Using the mouse-human numbers, we estimate that frogs and humans are 300 million years apart. That means that frogs and mice are separated evolutionarily by 240 million years.

b) Two duplications must have occurred to lead to alleles B and C in humans. Because the C allele has 300 bases different from A the A-C duplication occurred 30 million years ago and has undergone base changes over that time period. The B allele has fewer differences compared with C so was duplicated and evolved more recently than C. Using our clock rate, it arose by duplication 1 million years ago.

c) The duplication of B resulted in a copy on a different chromosome (or far enough away on the same chromosome so that it is unlinked). The duplication could have been a transposition type of event. Because C is a tandem copy, it could have arisen by a misalignment of repeated sequences around the original copy followed by a crossing over.

24-14. The plasmid could have been introduced (via natural transformation or conjugation- see chapter 13 for more information on these processes) from another species in the wild.

24-15. This aberrant phylogeny of the glucose-6-phosphate isomerase could indicate that this particular gene was introduced from a different species.

24-16. Chromosome mutations have the potential for much more dramatic effects. For example, regulatory regions of one gene get juxtaposed to regions of other genes; several genes can be lost by deletion.

24-17. a) exons b) genes

24-18. a) In vertebrates there are four *HOX* gene family clusters compared with one cluster in *Drosophila*, for example. This fits well with the tetraploidization hypothesis since the two doublings should lead to four copies of the genes.
b) The variation within the four gene family clusters could be the result of duplications within the gene family in certain lineages.

24-19. LINES or SINES can mediate genome rearrangements by unequal crossing-over between elements or can contribute regulatory elements adjacent to a gene.

24-20. The size of dinucleotide repeat sequences can change by unequal crossing-over between the repeated sequence within the element. (There is also some variation that can occur by slippage during DNA replication)

24-21. a) SINES or LINES
b) centromere satellite DNA (Although we don't know exactly the function, it appears to be needed for proper centromere function in higher eukaryotes.)

24-22. a) 2 b) 4 c) 1 d) 3

Notes

Notes

Notes

Notes

Notes

Notes

Notes

Notes